STUDIES IN THE
HISTORY OF MATHEMATICS

Studies in Mathematics

William Aspray
Charles Babbage Institute

Martin Davis
New York University

Jean Dieudonné
Paris, France

Harold M. Edwards
New York University

Jeremy Gray
The Open University

Thomas Hawkins
Boston University

Dale M. Johnson
MITRE Corporation

Jesper Lützen
Odense University

Herbert Mehrtens
Technische Universität Berlin

Gregory H. Moore
McMaster University

Studies in Mathematics

Volume 26

STUDIES IN THE HISTORY OF MATHEMATICS

Esther R. Phillips, editor
Lehman College
The City University of New York

Published and distributed by
The Mathematical Association of America

© 1987 by
The Mathematical Association of America (Incorporated)
Library of Congress Catalog Card Number 87-060581

Complete Set ISBN 0-88385-100-8
Vol 26 ISBN 0-88385-128-8

Printed in the United States of America

Current printing (last digit):
10 9 8 7 6 5 4 3 2

CONTENTS

Contents

INTRODUCTION

The subjects of the articles in this volume are taken only from modern and contemporary mathematics. Space considerations and the assumption (on the part of the editor) that its potential readers would be mainly professional mathematicians—most of whom teach at the college or university level—have dictated the omission of articles treating the earlier history of mathematics and its development in other cultures. To be sure, a fair number of mathematics' instructors are interested in mathematics in antiquity or in non-Western cultures; but the readily available material here, especially in the case of mathematics in antiquity, seems adequate both for teaching the history of mathematics and for the introduction of historical topics into the classroom. In contrast, there appears to be a greater need and demand for historical information related to the teaching of advanced topics and to the early development of the major fields of contemporary research. What is more difficult to justify is the omission from this volume of many equally important and interesting topics from the more recent history. Were personal tastes the sole criterion for inclusion, there would (at least) have been articles on the theory of the integral and trigonometric series and on Riemann surfaces (besides several others). Moreover, there would have been contributions from other historians of mathematics whose work I greatly admire.

Thus, there has been no attempt to put together a "balanced" (with respect to subject) collection of articles. Indeed, this volume

can best be described as a collection of articles, not generally
related to one another either by subject or philosophy, whose
purpose is to provide a sample of recent scholarship in the history
of mathematics and to suggest additional sources for further study
and exploration. The individual contributions are, by and large,
typical of current research on "modern" topics: each relies on
original published sources and most have made significant use of
archival materials, interviews, or other sources not always accessible or even familiar to the public. No author has attempted to write
a sweeping survey of a particular field; indeed most of the contributions analyze a significant "event" or development in the history
of mathematics. The authors themselves are representative of the
range of specialists who have made significant contributions in
recent years to the history of modern/contemporary mathematics.
Most may be characterized as specialists in the history of mathematics, in that the majority of their publications are in this area;
several (Davis and Dieudonné), however, are distinguished
mathematicians with long-standing interests in the history of
mathematics.

Until the publication in 1972 of Morris Kline's *Mathematical
Thought from Ancient to Modern Times* (Oxford University Press),
the topics covered in the general (English language) histories of
mathematics ranged mainly from mathematics in antiquity to what
may be called the beginning of modern mathematics, the latter
marked by Cauchy's attempts to rigorize the calculus and the
discoveries of non-Euclidean geometries. With the possible exception of the latter two topics, the treatment of nineteenth and
twentieth century developments was, generally, sketchy. These
earlier texts were intended mainly for classroom use, and the
inclusion of additional topics would have resulted in unacceptably
bulky volumes. More important, the study of contemporary topics
requires a greater mathematical sophistication and knowledge than
can be realistically expected from the majority of undergraduates.
(Of these earlier books, Carl Boyer's *A History of Mathematics* and
Howard Eves' *An Introduction to the History of Mathematics* can be
recommended.) Kline's monumental text (more than 1200 pages)
was the first and (to my knowledge) the only instance of a general
historical account in which a serious attempt was made to survey

major developments that took place in the latter half of the nineteenth century and, in some cases, in the early decades of this century. This work has been very widely read and cited, and it went a long way toward providing information on more advanced subjects. Yet, as the author reminds us in the Preface, it is a *survey*: "Euler's works fill some seventy volumes, Cauchy's 26 volumes, and Gauss's 12 volumes... a one volume work cannot present a full account." A reader who wanted, say, a fuller account of the origins and early development of group theory, would have been disappointed to find that Professor Kline was unable to provide many suggestions for further reading. The only recently published related article (cited on p. 770) is on Galois theory. Although *expository* papers on group theory were not difficult to find in the early 1970's, there were no adequate, readily available historical accounts (in English) available at that time. The only complete and reliable sources of information were the original papers and archival materials. Indeed, Professor Kline's book undoubtedly became (as its author hoped it would) the starting point for serious historical investigations: the absence of a "full account" of the development of a favorite topic was enough to drive some readers to the sources and, in some cases, to try to write such a full account.

During the mid to late 1970s interest in the history of mathematics on the part of "working" mathematicians—mainly at colleges and universities—increased, and various efforts were undertaken to make more information accessible to the mathematical community. In 1974 the journal *Historia Mathematica* (sponsored by The International Commission on the History of Mathematics) began publication under the direction of Kenneth O. May. Until 1986 the journal was published in North America (first in Canada, later in the U.S.A.), and the majority of its contributions were written in English. Almost from the time of its founding up to 1986, an ever increasing number of articles, queries, notes, etc., received by the editors, came from members of departments of mathematics. It may be added that during this period the number of papers on "modern" topics, submitted to the journal for publication, grew steadily; at the same time there was also a corresponding increase in the publication of historical monographs on modern/contemporary subjects, written for (and occasionally by)

professional mathematicians as well as for and by historians of
mathematics. Moreover, one can not have failed to notice that in
the last decade or so many good examples of "historical exposition"
(if not always, strictly and narrowly speaking, of history) have been
featured in *The American Mathematical Monthly*. The references
cited by the contributors throughout this volume give further
demonstration of the significant growth in recent years of the
number of publications in the history of modern and contemporary
mathematics. Yet, despite this growing number of publications,
there remain significant topics that have received scant attention
from historians, while others that may have already received extensive coverage continue to attract the attention of scholars. We
hope, therefore, that this collection will be a welcome addition to
this growing literature.

Each article in this volume contains a separate list of references.
Some primary sources, mainly of interest to historians, have been
omitted, and a few of the secondary source listings have been
annotated. These references, if not themselves complete, contain a
fairly complete set of "pointers" to more extensive lists of references on the topics of the articles contained in this volume. An
excellent source for publications in virtually every topic of the
history of mathematics is the recently published volume, *History of
Mathematics from Antiquity to the Present*, a selected bibliography,
edited by J. W. Dauben (New York: Garland, Inc. 1985). Recent
issues of *Historia Mathematica* and the *Archive for History of
Exact Sciences* will provide additional information.

Brief summaries of the articles are given below. Their order
coincides with the order in which the articles appear in the volume.
Beyond the fact that a few contiguously placed articles have
slightly overlapping subjects, there is no particular construction or
algorithm that would produce this ordering.

Richard Dedekind's introduction (beginning in 1871) of the
concept of "ideals" into algebraic number theory was perhaps the
most decisive step in the development of the modern set-theoretic
approach to the formulation of mathematical concepts. Harold M.
Edwards describes the context of this work and how it prevailed
over a competing theory, put forward by Kronecker, that did

"exactly what Dedekind set out to do—that is, it generalizes Kummer's theory from ideal factorization over cyclotomic fields to arbitrary algebraic fields...." Dedekind had stressed the relation between ideals and the "cut" concept, asserting that "an ideal number is 'concretely' represented by the set of all *actual numbers* that it is greater than...." Kronecker, who objected to the use of a completed infinite, offered his alternative approach.

Wilhelm Killing's choice of Weierstrass to be his thesis adviser at the University of Berlin was remarkable for the fact that Killing, from the outset, had been drawn to geometry; indeed, Killing had described his dissertation as an attempt to give a geometric interpretation to Weierstrass' theory of elementary divisors. Thomas Hawkins describes Killing's application of Weierstrassian techniques to problems in the foundations of geometry. These techniques involved an "exhaustive analysis of all algebraic possibilities" and ran counter to the characteristic algebraic, "generic" analysis of that period. Killing's theory of general space forms, drawing heavily from the ideas of Riemann and Helmholtz, are analyzed as the background to a remarkable series of papers, published by Killing in 1888, on Lie Algebras.

By 1830 J. Bolyai, K. Gauss, and N. Lobachevskii had made their (independent) discoveries of non-Euclidean geometry; but, according to Jeremy Gray, mathematicians were "entitled to be sceptical, because of the obscurantism inherent in the trigonometry of Lobachevskii and Bolyai," and it was not until the 1850s and 60s that their work began to be accepted by mathematicians. In his paper Gray describes "what geometry meant to the various protagonists," the delay in the acceptance of their work, and why so many distinguished mathematicians misunderstood their ideas.

The early development (1907–1910) of the topologist L. E. J. Brouwer is described in the article by Dale M. Johnson. After completing his thesis on the foundations of mathematics, Brouwer began a critical examination of A. Schoenflies' topology. This in turn led Brouwer to investigate the field of topology directly. The most important discovery made by Brouwer in this period was the concept of mapping degree, which became the principal tool in his topological work.

The emergence of first-order logic from richer forms of logic—found in the work of Hilbert and Skolem—and how second-order and infinitary logical notions functioned in the work of earlier logicians are discussed in Gregory H. Moore's article. Despite its inability to express the categoricity of the natural and real numbers, first-order logic became, after the publication of Gödel's papers on the completeness of first-order logic and the incompleteness of second-order and ω-order logics, the basis for mathematics.

In the course of solving what Hilbert had characterized as the principal problem of mathematical logic Alan Turing, in the mid 1930s, developed an abstract theory that suggested a far broader conception of computation than had hitherto been considered. Turing's definition of the universal computing machine and the influence of his ideas on the development of the electronic stored program digital computer in the 1940s and 50s are the subjects of the paper by Martin Davis.

The efforts of John von Neumann to have a computer installed at the Institute of Advanced Studies, Princeton are used by William Aspray to examine the relation between mathematics and computing in the decade 1945–55. Von Neumann was one of the first to appreciate this relationship and did much to advance it through his research in automata theory, computer design, numerical analysis, and mathematical physics, and through his promotion of scientific applications of the IAS computer.

Herbert Mehrtens' contribution describes the attempts made by Ludwig Bieberbach (whose well known conjecture was proved by Louis de Branges in 1984) to create a "German Mathematics." Central to the presentation are the political and sociological analyses of Bieberbach's attraction to Naziism and his extreme view of mathematics (already formed during the days of the Weimar Republic) in which "modern," "formal" mathematics, as exemplified in the work of Hilbert, was perceived to be destructive of mathematics itself. Bieberbach's struggle to bring about a joint political and mathematical revolution in Germany and the reactions of mathematicians, both in and outside Germany, are the subjects of this paper.

Jesper Lützen traces the application of the technique of separation of variables from the eighteenth century, when it first appeared in connection with the solution of simple partial differential equations, to the publication in the 1830s of the Sturm-Liouville Theory, which arose by applying the technique to equations with variable coefficients. Fourier series, for example, were first introduced in this way; more complicated boundary conditions led to the introduction of a variety of special functions.

Jean Dieudonné's article surveys the Italian school of algebraic geometry, which began in the early 1860s with L. Cremona's investigations of the structure of the most general birational transformations of the projective plane onto itself. Until 1900, which marks the end of the period under consideration, algebraic geometry was dominated by "two fundamental notions": birational correspondences and divisors. The article describes the contributions of the German school (made by R. Clebsch and M. Noether) to linear series (an interpretation of divisors); the beginnings of the theory of algebraic surfaces; linear systems of curves on algebraic surfaces; and the joint papers by Castelnuovo and Enriques [1900].

ESTHER R. PHILLIPS

DEDEKIND'S INVENTION OF IDEALS*

Harold M. Edwards

The present-day formulation of ideal theory is quite standardized. A subset of a commutative ring with unit is called an *ideal* if it is closed under addition and under multiplication by ring elements. The *product* of two ideals of a ring is the ideal consisting of all elements of the ring that can be written as sums of products $a \cdot b$ in which a is in the first ideal and b is in the second. The ring itself is an ideal and every ideal A of a ring R has the trivial factorization $A = RA$. An ideal A is called *prime* if it has no nontrivial factorizations, that is, if $A = BC$ implies $B = R$ or $C = R$. Under suitable conditions on R, *every ideal can be written as a product of prime ideals* and this representation is unique, up to the order of the factors.

This is almost exactly the way that Dedekind himself formulated the theory. He was dealing specifically with the ring of integers of an algebraic number field and did not use the term "ring," which was introduced later by Hilbert, and instead of speaking of a "subset" of the ring of integers he spoke of a "system" (*System* in

---

*This article was originally published in the *Bulletin of the London Mathematical Society* (volume 15, 1983, pp 8–17). The original version includes the sources of all the quotations, as well as the quotations themselves in their original languages.

German) of integers. Except for these minor differences, Dedekind's presentation of the theory in the supplements to the 3rd and 4th editions of Dirichlet's lectures on number theory (in 1879 and 1894) exactly fits this description.

It should be seen as a great triumph for Dedekind that his formulation of the basic concepts has won such universal acceptance. For him, the way in which one formulated the basic concepts of a mathematical theory was of the utmost importance, and his version of ideal theory was the result of the most careful deliberation, much in the way that Leibniz's philosophy assigned great importance to mathematical notation and the Leibniz notation for the calculus was the result of years of effort. And just as the success of Leibniz's notation is evident in today's notation for differential forms, the success of Dedekind's set-theoretical approach to the fundamental concepts is evident in today's approach to all of modern algebra and, indeed, today's approach to the foundations of mathematics in general.

The success of Dedekind's ideas has been so great that modern mathematicians unaware of the history have no idea how innovative they were. Pierre Dugac has rightly called Dedekind's first publication of his theory of ideals the "birthplace" of the modern set-theoretic approach to the foundations of mathematics. In order to appreciate how new it was, one must, of course, investigate the context in which it was created. This will also shed light on the considerations which led Dedekind to formulate the theory in the way that he did, and hence will give some insight into the reasons for the success of his ideas.

First, it should be said that there is no necessity whatsoever to formulate the theory this way. Kronecker's theory of divisors does exactly what Dedekind set out to do—that is, it generalizes Kummer's theory of ideal factorization from cyclotomic fields to arbitrary algebraic number fields—and it uses no "ideals" at all. Moreover, Dedekind's ideas aren't even sufficient. As I tried to show in my paper [Edwards, 1980] on the genesis of ideal theory, the really crucial idea is that of an "integer" in an algebraic number field. Without that—that is, without the correct determination of the *ring* to which the ideal theory is to apply—Dedekind's ideas are of no help. With that, the theory can be developed in at least three quite distinct ways, one of them Dedekind's, one of them Kronecker's, and the third one a fairly automatic generalization of Kummer's original approach.

In fact—and this shows how far Dedekind's formulation is from being the obvious one—this third approach following Kummer is essentially the one Dedekind himself took in his original publication of the theory in 1871 in the 2nd edition of Dirichlet-Dedekind. In this version, the now-familiar definition of ideals is used, to be sure, but the notion of the *product* of two ideals occurs only at the very end of the theory. In its stead, the basic role is played, as it was in Kummer's theory, by the notion of the *divisibility* of one ideal by another, which is defined by certain congruence conditions; the theorem on unique factorization states that if two ideals are divisible by the same prime ideals with the same multiplicities, then they are equal. (The definition of the multiplicities involves, from Dedekind's point of view, technical difficulties which he was eager to remove.) Thus, far from being the most obvious and direct approach to the subject, Dedekind's final version was one he himself arrived at only after years of revision and reconsideration.

This last fact also shows that while Kummer's theory of ideal factorization for cyclotomic integers was, of course, the inspiration for Dedekind's work, his final result represented in some ways a *rejection* of Kummer's ideas. There are three specific objections to Kummer's theory which Dedekind expresses and which shed light on the philosophy of mathematics from which his ideas sprang.

The fundamental concept of Kummer's theory is that of an *ideal prime factor* of cyclotomic integers.[1] Kummer defines what it means to say that a given cyclotomic integer is "divisible μ times" by a given ideal prime factor, but he never says what an ideal prime factor *is*. Moreover, he occasionally speaks of *ideal numbers*, rather than ideal prime factors, and he never says at all what an ideal number is. (One can infer from his use of the notion that by an ideal number he meant essentially a formal product of ideal prime

---

[1]Let λ be a prime (Kummer's notation) and let α be a primitive λth root of unity. A cyclotomic integer—Kummer called them "complex numbers formed from α"—is a number which can be expressed as a polynomial in α with integer coefficients. Computations with cyclotomic integers are easy to perform using the basic relation $1 + \alpha + \alpha^2 + \cdots + \alpha^{\lambda-1} = 0$. For details, see my book *Fermat's Last Theorem*.

factors.) This state of affairs was extremely unsatisfactory to Dedekind, who thought of *identifying ideal numbers* (and, in particular, ideal prime factors) *with the set* (system) *of integers that they divide*. He then found—not without considerable difficulty, for it is a difficult theorem—that such subsets are characterized by the fact that they are closed under addition and multiplication by ring elements. "This situation naturally led me to found the entire theory of numbers of the domain [of integers in a number field] on this simple definition [of "ideal"] *delivered of all obscurity and the admission of ideal numbers*" (emphasis added) [Dedekind, 1932, 271–272]. Thus Dedekind's formulation of the theory was a reaction to what he regarded as inadmissible obscurity and vagueness in Kummer's definition of ideal numbers.

Another aspect of Kummer's formulation that Dedekind rejected was that it was too specific. Kummer gave explicit rules for determining the ideal prime factors of a given cyclotomic integer. Dedekind saw the weakness of an approach that used specific properties of the cyclotomic integers and could not be applied to other number fields. "I am of course far from wanting to compare the value of [the development of ideal theory] with that of Kummer's original creative idea, but, apart from the aesthetic satisfaction of knowing that the same simple law according to which numbers are composed of primes applies in the form of the theory of ideals to any field, the knowledge of the general theory has the great practical advantage that one is not obliged to return in each special case to the first principles and that the working out of the special laws is simplified to an extraordinary degree" [Dedekind, 1932, 419].

The third way in which Dedekind's work represents a negative reaction to Kummer's is less subtle. Kummer's work contained mistakes. They are rare, and they do not affect the main conclusions of his papers, but naturally any doubt about the correctness of one's conclusions is the antithesis of good mathematics. And the source of Kummer's problem was his suggestive terminology. Dedekind politely put it this way: "Kummer never defined the ideal numbers themselves but only divisibility by them.... Although this introduction of new numbers is entirely legitimate, one may well fear that the chosen mode of expression in which one speaks of determined ideal numbers and their products, and the presumed analogy with the theory of rational numbers, may lead one to premature conclusions, and then on to inadequate proofs—and, in fact, this outcome has not always been avoided" [Dedekind, 1932, 268].

So the inspiration for Dedekind's set-theoretic approach did not come from Kummer. Nor, in my opinion, did it come from Gauss, under whose direction Dedekind wrote his dissertation (on Eulerian integrals). Of course Gauss's appreciation of the role of generalization in the development of mathematics, and his unerring judgement as to what topics would prove fruitful, had a profound influence on the mathematics of that period and therefore on Dedekind. However, the set-theoretic approach is far removed from anything we know of Gauss's work. Anyone familiar with Dedekind's biography will be well aware that the decisive influence on his mathematical outlook was rather his association with the great G. Lejeune-Dirichlet.

When Gauss died in early 1855, just as Dedekind was beginning his period as a Privatdozent in Göttingen, Dirichlet came from Berlin to replace Gauss in Göttingen. His presence made of Dedekind, in his own words, a "new man." Paul Bachmann, who was a student in Göttingen at the time, recalled in later years that he only knew Dedekind by sight because Dedekind always arrived and left with Dirichlet and was completely eclipsed by him. Dedekind himself, in a letter of July 1856 that was recently published, describes how great a difference Dirichlet's presence made. "What is most useful to me is the almost daily association with Dirichlet, with whom I am for the first time beginning to learn properly; he is always completely amiable toward me, and he tells me without beating about the bush what gaps I need to fill and at the same time gives me the instructions and the means to do it. I thank him already for infinitely many things, and no doubt there will be many more" [Scharlau, 1981, 35].

Hermann Minkowski, in a memorial lecture about Dirichlet spoke of the "other Dirichlet Principle" (*the* Dirichlet Principle being, of course, the notion in function theory to which Riemann gave that name) which he described as being the principle of "overcoming problems with a minimum of blind calculation and a maximum of perceptive thought" [Minkowski, 1911, 460–461]. It is this other Dirichlet principle that I see as having been the primary influence on Dedekind's philosophy of mathematics and on the development of Dedekind's mathematical style.

Of course Dedekind's own personality determined the way in which this influence would act. Kummer, too, was strongly influenced by Dirichlet, but, since his personality was so different, the influence acted in a different way. Basically, Kummer was a *computer* and Dedekind was an *expositor*. I don't mean, of course, that Kummer was *only* a computer or that Dedekind was *only* an expositor—both were mathematicians of wide-ranging abilities and great accomplishments. I only mean to give an indication of what seems to me a clear difference in their attitudes and ways of thinking about problems.

That Dedekind took his teaching very seriously, even at this early stage of his career, is clear. Dirichlet, in a letter recommending appointment of Dedekind to the Polytechnikum in Zürich, described him as "an exceptional pedagogue," and once Dedekind arrived in Zürich (he got the job) his teaching was an integral part of his mathematical thinking. Another proof of Dedekind's calling as an expositor is his publication in 1863 of the first edition of "Dirichlet's" lectures on number theory. (Although the book is assuredly *based* on Dirichlet's lectures, and although Dedekind himself referred to the book throughout his life as Dirichlet's, the book itself was entirely written by Dedekind, for the most part after Dirichlet's death.) The importance of Dedekind's "supplements" to the later editions of this book should not be allowed to obscure the importance of the book itself, which was surely the most important book on number theory after Gauss's *Disquisitiones arithmeticae*. Shortly after it appeared, Kronecker mentioned it in the following way in a paper presented to the Berlin Academy:

> I should remark, in addition, that a similar derivation is also to be found in one of the very worthwhile supplements which Herr Dedekind added to his extremely commendable edition of Dirichlet's lectures on number theory, which is so skillfully and carefully executed [Kronecker, 1929, Vol. 4, 232].

This demonstrates, I think, the importance of Dirichlet's influence on Dedekind's mathematical formation, and on his attention to matters of exposition and "perceptive thought" as opposed to "blind calculation." Another major influence was the work of Riemann. Dedekind himself speaks of this influence in a letter he wrote in 1876 to R. Lipschitz:

> My efforts in number theory have been directed toward basing the work not on arbitrary representations or expressions but on simple foundational concepts and thereby—although the comparison may sound a bit grandiose—to achieve in number theory something analogous to what Riemann achieved in function theory, in which connection I cannot suppress the passing remark that Riemann's principles are not being adhered to in a significant way by most writers—for example, even in the newest works on elliptic functions. Almost always they mar the purity of the theory by unnecessarily bringing in forms of representation which should be results, not tools, of the theory [Dedekind, 1932, 468–469].

Dedekind's aversion to basing anything in mathematics on forms of representation, which is so clear in this quotation, is expressed in another place in the correspondence with Lipschitz, where he says that in the exposition of ideal theory he himself published in the *Bulletin des Sciences Mathématiques*, in the interest of brevity, he marred (*ich verunziere*) the theory by using explicit representations of fields in the form $\mathbb{Q}(x)$, where x satisfies an algebraic relation over $\mathbb{Q}$. To Dedekind, the field is what is important, and the arbitrarily chosen representation $\mathbb{Q}(x)$ is to be avoided.

This Riemannian rejection of forms of representation and the use instead of characteristic properties as foundational concepts is closely related to that "other Dirichlet principle" and can in fact be viewed as another influence of Dirichlet on Dedekind, this time through the medium of Riemann. As Siegel has observed, Riemann was a computer of astonishing power and skill, but his published work always emphasized conceptual thinking and avoided formulas and "blind calculation."

In considering possible influences on the development of Dedekind's set-theoretic approach, mention should be made of the work of Galois. Galois has been widely praised for the modernity and abstractness of his thinking, and it is known that Dedekind studied Galois's work very carefully during the years 1856–1858. Nonetheless, I am not able to see an influence here. Certainly the effort to understand and to give clear expositions of Galois's theory

(Dedekind was the first to teach Galois theory at a German university) was a formative experience for Dedekind, but Galois showed no reluctance to base his formulation of Galois theory on the arbitrary choice of a Galois resolvent to represent the splitting field, and there appears to be no hint of the set-theoretic approach in his work.

But there is more to the set-theoretic approach than conceptual thinking. Dirichlet, Riemann, and Galois were all conceptual thinkers—and so, no doubt, were Gauss and many others—but this does not imply that they described the basic concepts in terms of sets. In fact, they most certainly would *not* have described basic concepts in terms of sets because of the classical doctrine which excluded any use of a completed infinite. As Gauss said in an oft-quoted letter (July 12, 1831) to Schumacher, "But concerning your proof, I protest above all against the use of an infinite quantity as a *completed* one, which in mathematics is never allowed. The infinite is only a *façon de parler*, where one would properly speak of limits" [Gauss, 1900, 216].

It must be asked what motivated Dedekind to defy this doctrine to the extent of *defining* his ideals as infinite sets, and to regard these infinite sets as completed entities. In this he was not following Dirichlet, and it seems unlikely that Dirichlet, had he been alive, would have followed him. The best indication I know of Dirichlet's views in such matters, comes, ironically, from Dedekind's introduction to *Was sind und was sollen die Zahlen?*: "...Seen in this way it seems perfectly obvious and by no means new that every theorem of algebra and higher analysis, no matter how remote, can be expressed as a theorem about the natural numbers, an assertion that I heard repeatedly from the lips of Dirichlet" [Dedekind, 1932, 338]. One can be certain that Dirichlet did not regard the natural numbers as a completed infinite and that his assertion here is much akin to Gauss's about the infinite—in Dirichlet's case, algebra and higher analysis—being a *façon de parler*.

Dedekind's contrary view—his use of completed infinites as basic concepts—goes back to his earliest writings. As Dugac has observed, already in his 1857 memoir, *Abriss einer Theorie der höheren Kongruenzen*, Dedekind described addition modulo a prime as addition of equivalence classes, that is, as addition of infinite sets regarded as entities. However, in all of elementary algebra, as in this example, the infinite sets involved are of a very tame type: they can be described explicitly in a way that is acceptable to even the most finitistic philosophy of mathematics. What prompted Dedekind to fly in the face of the doctrine against completed infinites could not have been anything in algebra: it must have been something from analysis.

In fact, there had been trouble brewing for quite a while in analysis. Consider, for example, the function concept. One of Dirichlet's greatest achievements was his proof in 1829 that the Fourier series of a function converges to the function for an extremely general class of functions, namely, all functions "which have a finite number of maxima and minima" as it was expressed at the time or which are "piecewise monotone" as we would express it today. What *is* such a function—how is it to be described? Surely not by a formula. The whole point of the theorem was to show that the functions to which Fourier analysis applies far transcend the functions given by simple formulas. "Functions that are given by a graph" was the way general functions were described in those days. But what is a graph but a subset of $\mathbb{R} \times \mathbb{R}$? And what is a subset of $\mathbb{R} \times \mathbb{R}$ but a completed infinite? Set theory is what is left after you reject all formulas.

It was just such considerations of very general functions and the convergence of Fourier series that led Cantor in the 1870s and 1880s to his bold advancement of the theory of infinite sets. For Dedekind the impulse to overthrow the doctrine against the completed infinite came from a different direction—the need to elucidate the idea of a real number. Dedekind tells us the exact date—November 24, 1858—on which the idea occurred to him to describe real numbers in terms of what we know today as *Dedekind cuts*. What prompted his thoughts in this direction was, characteristically, his teaching. This was his first year in Zürich and it was the first time he had taught differential and integral calculus. In his attempts to formulate an exposition of the fundamental concepts of the calculus for his students which he himself found acceptable, he found the existing expositions to be of no help and decided to do it on his own. The result was his description of real numbers in terms of the "cuts" of the rational numbers that they generate: A real number—whatever it is—effects a division of the rationals into two sets, those less than it and those greater than it. Dedekind turned the tables and took such a division of the rationals to *be*, or at any rate to *represent*, a real number.

The analogy between cuts and ideals is obvious. An ideal number in a ring is "concretely" represented by the set of all *actual* numbers of the ring that it divides. A real number is "concretely" represented by the set of all *rational* numbers that it is greater than (or by those it is less than). Here the use of the term "concrete" would scarcely be acceptable to anyone—Gauss, for example—who rejected the completed infinite. Still, it was "concrete" to Dedekind, and he used this term without apology, unless his delay of the publication of the idea until 1872 can be regarded as an acknowledgement that there might be strong objections.

Kronecker did object:

> These considerations oppose, it seems to me, the introduction of Dedekind's notions of modules, ideals, etc., as well as the introduction of various new concepts that have been used lately (in the first instance by Heine) in the attempt to understand and to give foundations to the "irrational" in full generality. The *general* concept of an infinite series itself, for example a power series, is in my judgement permissible only with the reservation that in each particular case the arithmetical rule by which the terms are given satisfies, as above, conditions which make it possible to deal with the series as though it were finite, and thus to make it unnecessary, strictly speaking, to go beyond the notion of a finite series.

This objection was in a footnote to a paper published in 1886 [Kronecker, 1929, Volume 3, 156].

Dedekind replied with a footnote of his own, published in *Was sind und was sollen die Zahlen?* (1888):

> In what way the determination [whether a given element is in a given set] comes about, or whether we know a way to decide it, is a matter of no consequence in what follows. The general laws that are to be developed do not depend on this at all. They hold under all circumstances. I mention this expressly because Herr Kronecker a short while ago sought to impose certain restrictions on the free introduction of concepts into mathematics, restrictions which I do not regard as justified. To discuss this in detail will be called for, however, only after that eminent mathematician publishes his reasons for asserting the necessity—or even simply the usefulness—of these restrictions [Dedekind, 1932, 345].

I award that round to Dedekind.

On the whole, the objections to Dedekind's ideas were surprisingly few, and the victory of his point of view was surprisingly swift and complete. One of the few indications of resistance other than Kronecker's is to be found in Dedekind's correspondence with Lipschitz. It seems that Lipschitz advised Dedekind *against* citing the analogy between ideals and Dedekind cuts in his exposition of ideal theory for the *Bulletin des Sciences Mathématiques*. Evidently Lipschitz felt that the Dedekind cut notion was not well accepted (not, however, because of its use of a completed infinite) and that this would not be seen as a point in favor of ideals. Dedekind's reply shows that Lipschitz had struck a nerve. The normally patient, mild, and altogether genial Dedekind suddenly became dogmatic, insistent, and almost out of conrol (only a *portion* of this letter is published in Dedekind's *Werke*—the whole letter is in the Göttingen archives) as he argues for the importance of the cut concept and for the necessity of some such concept for the proper foundation of analysis. This also demonstrates, if a demonstration is needed, the close relationship between ideals and Dedekind cuts. As I said above, the infinite sets needed in algebra are very tame; Dedekind would not have had the slightest difficulty in defining ideals in a way that met the most stringent restrictions on the use of the infinite, but he was not willing to take any steps in this direction because he knew full well that the same could not be done for his cuts.

There are several reasons for the success of Dedekind's ideas among his contemporaries. One was surely the publication in 1882 of his epoch-making paper with Heinrich Weber, which applied his ideal theory with such stunning success to the theory of Riemann surfaces (algebraic function fields of one variable). Here for the first time the Riemann-Roch theorem was proved rigorously under the most general assumptions using purely algebraic means—Dedekind's ideals and other ideas adapted from his work on number theory.

Another reason for the success of Dedekind's ideas was the marvellous clarity with which he presented them. He always assigned great importance to good exposition, and his writing shows it. By contrast, his rival, Kronecker, published a major work, his

Kummer *Festschrift*, at exactly the same time that the paper of Dedekind-Weber appeared. No one doubted the importance of Kronecker's ideas, but it seems that no one could read them. The contrast could hardly have been more damaging from the point of view of winning allies for Kronecker's ideas on the foundations. With Kronecker, even his objectives were obscure, much less the question of whether or to what extent he achieved them. With Dedekind-Weber, the objective was one whose importance was universally recognized, and it was achieved with absolutely Euclidean economy and clarity.

The appearance of Cantor's revolutionary works on set theory in the 1880's may also have contributed to the acceptance of Dedekind's formulations. Cantor's use of the infinite was so much more aggressive than Dedekind's that mathematicians with scruples in such matters concentrated their fire on Cantor, not on Dedekind, and certainly not on Dedekind's number-theoretical works.

Dedekind's ideas also succeeded through a chance of circumstance that deserves to be mentioned. At the time Dedekind-Weber appeared, Weber happened to be teaching number theory at the University of Königsberg. Without doubt, Weber, who was the junior partner in the collaboration, would have done number theory at that time in an entirely Dedekindian way. One of his students was the young David Hilbert. Hilbert's early exposure to Dedekind's approach most probably was a help to him in his development into the foremost mathematician of his generation, and to this extent it is not mere chance that Dedekind's viewpoint came into ascendancy via Hilbert. On the other hand, talents like Hilbert's are rare, and without him things might have been very different.

The extent to which objections to the completed infinite were forgotten is astounding. Hilbert's proof of the existence of an integral basis in the *Zahlbericht* (1897) begins "We imagine all the integers of the field in a list." He then proceeds to pick out of his imaginary list an entry in which a particular term has the smallest possible positive value! I have no way of proving it, but I believe even Dedekind would have balked at such a sweeping use of the completed infinite, especially to prove a theorem which can so easily be proved constructively, as Dedekind always did.

Dedekind once wrote of Kronecker that "I have always been far from wanting to compare my *aurea mediocritas*, the strength of which lies in obstinate perseverance, with Kronecker's outstanding talent." In this essay I have tried to conceal my personal prejudice against Dedekind and for Kronecker, when it comes to questions about the foundations of mathematics. Even with this personal prejudice, however, I do not accept Dedekind's modest estimate of himself in comparison to Kronecker in the sentence just quoted. Kronecker's brilliance cannot be doubted. Had he had a tenth of Dedekind's ability to formulate and express his ideas clearly, his contribution to mathematics might have been even greater than Dedekind's. As it is, however, his brilliance, for the most part, died with him. Dedekind's legacy, on the other hand, consisted not only of important theorems, examples, and concepts, but of a whole *style* of mathematics that has been an inspiration to each succeeding generation.

REFERENCES

This list of works is merely a key to the quotations, not a bibliography. For a bibliography, see [Edwards, 1980]. A much more complete bibliography appears with the edition of Dedekind's "Bunte Bemerkungen zu Kronecker's *Grundzuge*," edited by H. Edwards, O. Neumann, and W. Purkert, in *Archive for History of Exact Sciences*, 27 (1982), 49–85.

Dedekind, R., 1932, *Gesammelte mathematische Werke*, vol. 3, Vieweg, Braunschweig. Reprinted by Chelsea Publ. Co., New York, 1969.

Dugac, P., 1976, *Richard Dedekind et les fondemonts des mathématiques*, Vrin, Paris.

Edwards, H. M., 1980, "The genesis of ideal theory," *Archive for History of Exact Sciences*, 23, pp. 321–378.

Gauss, C. F., 1900, *Werke*, vol. 8, Teubner, Leipzig.

Kronecker, L., 1929, *Werke*, vol. 4, Teubner, Leipzig. Reprinted by Chelsea Publ., Co., New York, 1968.

Minkowski, H., 1911, *Gesammelte Abhandlungen*, vol. 2, Teubner, Leipzig.

Scharlau, W., (Ed.), 1981, *Richard Dedekind 1831 / 1981*, Vieweg, Braunschweig/Wiesbaden.

NON-EUCLIDEAN GEOMETRY AND WEIERSTRASSIAN MATHEMATICS: THE BACKGROUND TO KILLING'S WORK ON LIE ALGEBRAS*

Thomas Hawkins

In 1888 a series of papers began to appear in *Mathematische Annalen* entitled "The Composition of Continuous Finite Transformation Groups [Killing, 1888a, 1888b, 1889, 1890]. The appearance of these papers must have raised some eyebrows because they seemed to constitute a major contribution to mathematics and yet the author, Wilhelm Killing, was a little-known, forty-one-year-old Professor at the Lyceum Hosianum in Braunsberg, East Prussia (now a part of Poland). The Lyceum was a Roman Catholic training center for future clergymen. Despite the unlikely background of the author, the papers did in fact constitute a major contribution to mathematics, a contribution as unexpected as it was extraordinary. It was in these papers that the entire theory of the structure of semisimple Lie algebras originated. Here we find the origins of such key concepts as the rank of an algebra, Cartan subalgebra, Cartan integers, root systems, nilpotent and semisimple

---

*This essay is a reprint, with minor changes, of [Hawkins, 1983]. I wish to thank the New York Academy of Sciences for kindly granting permission to reprint.

algebras, and the radical of an algebra, as well as fundamental results such as the theorem enumerating all possible structures for finite-dimensional simple Lie algebras over the complex field and a radical splitting theorem.

The purpose of this essay is to discuss how such an unlikely figure as Killing came to create such unexpected mathematics. As the title suggests, two factors principally determined the direction of Killing's research. The discoveries in non-Euclidean geometry and the concomitant speculations on the foundations of geometry formed the context of Killing's work. His contributions to the theory of Lie algebras were a by-product of his research program on the foundations of geometry. But when compared with contemporaneous work on non-Euclidean geometry and its foundations, Killing's work stands out as atypical. Since it is precisely the peculiar emphasis of Killing's research program that brings with it the algebraic problem of determining, in effect, all possible structures for Lie algebras, it is of considerable historical interest to seek to understand its basis.

In this connection, Killing's mathematical education is paramount. He received his mathematical training in the school of mathematics centered about Karl Weierstrass at the University of Berlin. As will be seen in what follows, during the decade (1867–1877) he spent in Berlin, Killing acquired the mathematical tools and general outlook on mathematics that oriented his geometrical research program in such an unusual manner.

To fully appreciate the magnitude of the impression Berlin made upon Killing it is necessary to go back a bit further in his life. Killing was born and raised in Westphalia. His aspiration as a *Gymnasium* student was to become a professor of mathematics at such an institution. The prospect of a university professorship apparently appeared too remote to take seriously. To prepare for a teaching career, Killing proceeded to the local university at Münster. The choice of Münster was an unfortunate one because at that time there were no mathematicians on the faculty. Mathematics courses were taught by an observational astronomer who openly admitted that his mathematical training was slight. The courses he gave were consequently elementary. In a sense, they catered to the demands of the students who, much to Killing's disgust, were primarily concerned with learning just enough mathematics to pass the examinations. Killing found himself forced to renew the practice he had found necessary as a *Gymnasium* student: self-study of works by distinguished mathematicians. For example, to supplement the "pablum" presented in the course on analytic geometry, Killing studied the treatises on this subject by Plücker and Hesse.

After two years of inferior education, Killing finally left Münster for Berlin. Coming from Münster, the impression Berlin made upon Killing must have been spectacular, for the University of Berlin was by then the leading center of mathematics in Germany and possibly in the world. The mathematics faculty, although of modest size by today's standards, consisted of Kummer, Weierstrass, Kronecker, and Fuchs: They were surrounded by a growing number of talented pre- and postdoctoral students. There was also a mathematics union run by students to further the dissemination of mathematical knowledge by means of sponsored lectures, problem-solving contests and the purchase of library books on mathematics. Instead of boring, elementary lectures, Killing could hear challenging lectures by Kummer, Weierstrass, and Kronecker coordinated so as to provide students with a solid background in up to date mathematics. In particular, there was the demanding cycle of lectures on analysis by Weierstrass, with their emphasis upon a rigorous, systematic development of analysis built upon an arithmetical foundation and without recourse to intuition or physical considerations. Finally, there was the famous mathematics seminar, run by Kummer and Weierstrass, in which students were brought to the frontiers of mathematics. (For further details, see the informative study by Biermann [1973].)

At Berlin, Killing was attracted above all to Weierstrass, as were many students. But Killing was at heart a geometer, whereas Weierstrass was an analyst in the tradition of Lagrange, Cauchy, and Jacobi. Kummer would have been a more logical choice for a doctoral dissertation advisor since 18 out of 31, or 58%, of the dissertations done under his direction were on geometrical topics. Nonetheless, Killing chose to work with Weierstrass. The dilemma of working with him and yet doing something geometrical was resolved by Weierstrass' theory of elementary divisors [1868], published the year following Killing's arrival in Berlin.

The theory of elementary divisors is concerned with necessary and sufficient conditions that one family of bilinear forms $A - \lambda B$ ($|B| \neq 0$) be transformable by means of linear variable changes into another such family. The theory is thus equivalent to the more familiar theory of canonical matrix forms, and in fact the elementary divisors of the characteristic polynomial $|A - \lambda B|$ are the divisors corresponding to the Jordan blocks of the Jordan canonical form for $A - \lambda B$. Weierstrass actually introduced this canonical form in his paper on elementary divisors and therefore prior to Jordan [Hawkins, 1977a].

The theory of elementary divisors represented Weierstrass' critical response to the prevalent tendency in eighteenth and nineteenth century algebraic analysis to reason vaguely in terms of a sort of "general case," according to which the algebraic symbols involved are regarded as assuming "general" rather than specific values [Hawkins, 1975, 1977a, 1977b]. This mode of reasoning brought with it a tendency to overlook, or ignore, potential limitations to the conclusions of a mathematical argument. This tendency was widespread and not limited to inferior mathematicians. Lagrange, Laplace, Jacobi, Hermite, and Cayley were among the practitioners of "generic" analysis. Within the context of the transformation of quadratic and bilinear forms, where "in general" the roots of characteristic polynomials are all distinct, Weierstrass rejected the tenability of such an attitude. An exhaustive analysis of all the algebraic possibilities was demanded by Weierstrass and facilitated by his theory of elementary divisors.

In Berlin the theory of elementary divisors was thus regarded as proving more than mathematical theorems. It demonstrated the desirability and feasibility of a more rigorous approach to algebraic analysis, one that did not shy away from the multitude of special cases that present themselves when the "general case" is abandoned. The following passage from a paper of 1874 by Kronecker exemplifies this attitude toward Weierstrass' theory. After referring to it, Kronecker wrote:

It is common, especially in algebraic questions, to encounter essentially new difficulties when one breaks away from those cases customarily designated as general. As soon as one penetrates beneath the surface of the so-called generality that excludes every particularity into the true generality, which includes all singularities, the real difficulties of the investigation are usually first encountered but, at the same time, also the wealth of new viewpoints and phenomena which are contained in its depths [Kronecker, 1874, 405].

It was in this spirit that Killing proposed to use Weierstrass' theory of elementary divisors to study, as his doctoral dissertation, pencils of quadric surfaces [Killing, 1872]. In homogeneous coordinates, such a pencil corresponds to a family of quadratic forms $A - \lambda B$ in four variables. The various geometrical possibilities can be systematically explored by first of all classifying according to the thirteen possibilities for the elementary divisors of the characteristic polynomial, and this is the approach taken by Killing. In the preface to his dissertation Killing explained that his intention was simply to provide a *geometrical interpretation* of his mentor's theory of elementary divisors. But, in effect, Killing also gave a geometrical interpretation to the Weierstrassian approach to mathematics: in geometrical investigations also one must systematically explore all the possibilities revealed by the analytical framework as viewed from the Weierstrassian standpoint rather than in "general" terms. This was not a widespread attitude among geometers of the period. As will be seen, Killing attacked the foundations of geometry in the same spirit as he had attacked quadric surfaces.

Killing received his doctorate in March of 1872, but remained in Berlin until 1878, teaching at the *Gymnasium* level. During the summer semester of 1872 the new doctor of mathematics participated in the mathematics seminar, and it was there that he heard Weierstrass present some lectures on the subject that was to become his life work: the foundations of geometry. Weierstrass was, of course, interested in foundational matters, but this subject was also a timely one because several events had combined to focus the interest of mathematicians and philosophers on non-Euclidean geometry and related foundational issues. The publication, during the mid-1860s, of Gauss' extensive correspondence with the astronomer Schumacher served to rescue the work of Lobachevsky from oblivion; for although Gauss carefully refrained from publicly expressing his favorable estimate of Lobachevsky's geometry his letters to Schumacher, published posthumously, conveyed his high opinion of Lobachevsky's work. Then in 1868 Beltrami showed that plane Lobachevskian geometry could be interpreted on a surface of constant negative curvature, thereby convincing sympathetic mathematicians that Lobachevsky's geometry did not contain hidden contradictions. In the same year, the philosophically oriented essays of Riemann [1868] and Helmholtz [1868] on the foundations of geometry appeared.

The Berlin Seminar of 1872 opened the door to a new mathematical world for Killing, the realm of non-Euclidean geometry. It will be helpful at this point to indicate the principal discoveries in this realm and their significance from Killing's perspective. The discovery of Lobachevskian geometry showed that the geometry of Euclid was not the only geometry both logically consistent and, in so far as could be determined, compatible with experience. The discoverers of Lobachevskian geometry (e.g., Lobachevsky, Bolyai, and Gauss) tended to regard it in absolute terms as exhausting the possibilities for geometry. That is, Lobachevskian geometry contained a parameter k, and for $k = \infty$ Euclidean geometry is obtained. Thus Lobachevskian geometry, including therewith the limiting case, embraced Euclidean geometry, and, in the eyes of its discoverers, all conceivable geometrical possibilities.

When Riemann, in his celebrated essay, suggested the possibility of a geometry of "finite space" corresponding to a manifold of constant positive curvature because the unboundedness of space was more certain that its infinitude, he therefore caused something of a sensation. Although, spherical geometry had long been familiar, Riemann was apparently the first to seriously consider this sort of geometry as an alternative to the geometry of Euclid. Whether Riemann actually intended to identify his geometry of finite space with spherical geometry is not certain, due to the vagueness of his primarily nonmathematical description, but most mathematicians, including Beltrami and Weierstrass, assumed Riemann was speaking of spherical geometry. Given this interpretation, Riemann's geometry of finite space was observed to violate the "axiom of the straight line," that two points *uniquely* determine a straight line. No sooner had the violation been assumed a necessary characteristic of a geometry of finite space (or of constant positive curvature) than Felix Klein and Simon Newcomb pointed out, independently, that there exists another geometry of finite space in which the axiom of the straight line does hold. Klein called it elliptical geometry.

Each new discovery thus revealed the limitations of the previous conceptions of the geometrical possibilities, a point that made a great impression upon Killing. The situation in geometry was analogous to what was occurring in the foundations of analysis, where earlier conceptions and intuitions regarding, for example, the properties of continuous functions were gradually being undermined by discoveries such as that of Weierstrass' example of a continuous nowhere differentiable function. Since Weierstrass presented his example to the Berlin Academy in July of 1872, it is likely that he also presented it in the same mathematics seminar where Killing had heard him lecture on the foundations of geometry. The analogy between developments in geometry and the theory of functions tended to reinforce Killing's attitude towards the foundations of geometry. In fact, Killing invoked the existence of functions such as Weierstrass' example to refute the arguments of one geometer who attempted to demonstrate that there existed exactly three possible geometries: Euclidean, Lobachevskian, and spherical [Hawkins, 1980, 314–315].

The discoveries in non-Euclidean geometry, reinforced by the discoveries in real analysis, convinced Killing that the science of geometry must be conceived in very general terms. For example, Killing concluded that it was impossible to completely capture spatial intuitions in an axiom system since diverse "models" (as we would say) can represent the same system. Euclid's system can be realized on the surface of zero curvature or by Klein's projective model (parabolic geometry). Nor could an axiom system uniquely capture the human experience of space. According to Killing, there were (at least) four geometrical systems compatible with experience: Euclidean, Lobachevskian, spherical, and elliptical.

Killing therefore concluded that the science of geometry is necessarily very general, its only requirement being logical consistency and completeness in the sense of embracing all the geometrical possibilities. Consequently:

> We regard all investigations of this type as branches of the same science and designate every individual possibility, with its further consequences, as a space form. Many may directly contradict experience, others may be very

unlikely, but they nonetheless all rest on the same foundations and exhibit in their proof procedures an unmistakable similarity   [Killing, 1880, 4].

In the spirit of Weierstrass, Killing proposed an exhaustive analysis of all the geometrical possibilities—all possible space forms—without recourse to the delimiting deceptions of intuition and experience, including thereby geometries that could be as counter-intuitive as continuous, nowhere-differentiable functions.

Among geometers, Killing's attitude was not at all typical. No one, including Killing, took seriously Riemann's vague hints that physical science might require a geometry corresponding to a manifold of variable curvature. Experience, the seemingly essential homogeneity of space, required a geometry corresponding to a manifold of constant curvature. Thus Riemann's greatest contribution to geometry was, in the view of Felix Klein, the concept of a manifold of *constant* curvature [Hawkins, 1980, 311, no. 40]. Such a manifold formed the analytical context for geometry and limited the possibilities to the known types of Euclidean and non-Euclidean geometries. Because the distinction between the local and global characteristics of a geometry was glossed over for a long time, the geometrical possibilities seemed limited to most geometers, although at the same time it must be remembered that non-Euclidean geometry was still (in 1860–1880) a radically new subject, a new mathematical world awaiting exploration and in this sense did not give the impression of being "limited."

Nonetheless, Killing's conception of the scope of geometry, of the geometrical possibilities, was considerably more far-reaching than that of his contemporaries. Only Riemann, with his idea of a manifold of variable curvature, entertained a view of geometry of comparable generality. But Killing's motivation was entirely different. Killing did not wish to transcend the limits of geometry as conceived by his contemporaries in the belief (such as Riemann held) that a more general conception of geometry might prove necessary to deal with physical reality. In arguments with his colleague R. von Lilienthal, Killing admitted that Euclidean geometry was the "only true" geometry, and he always regarded the well-known geometries of constant curvature as the most important because they corresponded to experience [Hawkins, 1980, 311]. But these sentiments did not diminish the need Killing felt to transcend experience and intuition in order to secure the foundations of geometry. This was of course precisely Weierstrass' attitude toward the foundations of analysis.

In their essays on the foundations of geometry both Riemann and Helmholtz made assumptions about the metrical properties of space. Riemann had argued tentatively that metrical relations were determined by a quadratic differential form; and Helmholtz, who approached geometry through the motions of rigid bodies, which included the assumption of some sort of a distance function left invariant by motions, argued that infinitesimal distances are expressible as a quadratic differential form, as Riemann had claimed. The treatment of these matters by both Riemann and Helmholtz was, however, lacking in clarity and rigor. It was probably because of this that Weierstrass, in his seminar lectures in 1872, called for a further exploration of possible metrics and the resulting geometries.

Killing went a step further and decided it would be best to begin without any assumptions of a metrical nature:

> Attempts to create a natural foundation for geometry have hitherto not been accompanied by the desired success. The reason for this lies, in my opinion, in this: just as geometry had to abandon the concept of direction in the sense stipulated by the parallel axiom, so also the concept of distance cannot be maintained as a general basic concept and therewith [geometry] must go far beyond the non-Euclidean space forms in the narrower sense [of Euclidean, Lobachevskian, spherical and elliptical geometry]   [Killing, 1884, iv].

Following Helmholtz, who was Professor of Physics at Berlin from 1870 onward, Killing approached the foundations of geometry through the motions of rigid bodies but without introducing metrical concepts. Killing's objective was to explore systematically and analytically the possibilities for his general space forms and, once this was achieved, then to consider the question of metrical properties.

The analytical starting point of Killing's theory of general space forms is the notion of an n-dimensional "manifold" of points $x = (x_1, \ldots, x_n)$ endowed with "m degrees of mobility." In the spirit of Riemann and Helmholtz, Killing concentrated on the infinitesimal aspects of the geometry and thus he dealt exclusively with infinitesimal motions $x \to x + dx$, where

$$dx_\rho = u^{(\rho)} dt, \qquad u^{(\rho)} = u^{(\rho)}(x).$$

Certain properties of these motions followed from generally accepted practices. Thus if (in abbreviated notation) $dx = u\,dt$ is a motion, so is $dx = \lambda u\,dt$. According to Killing, it is essentially the same motion if velocity is ignored. Likewise if $dx = u\,dt$ and $dx = v\,dt$ are motions, their "composite" $dx = (u + v)\,dt$ is another motion. This was the customary notion of the composition of two infinitesimal motions; the same parallelogram rule was followed as in the addition of forces. The composition of infinitesimal motions was thus taken to be commutative, as we would say nowadays. These generally accepted properties of infinitesimal motions were used by Killing to assert, in modern language, that the infinitesimal motions of a space form constitute a vector space. Killing assumed that this vector space had a finite dimension, m, which he called the degree of mobility of the space form. Killing's objective was to determine all possible space forms in n-dimensions with m degrees of mobility.

In the commonly accepted Euclidean and non-Euclidean geometries, one had $m = n(n + 1)/2$, but Killing of course, imposed no such restrictive condition on his space forms. The problem of exhaustively determining the possibilities was thus a difficult one, and although Killing did not really appreciate how difficult it would prove to be, he did sense the need to go beyond the traditional treatment of infinitesimal motions to make his problem tractable. In this connection he observed that although, traditionally, the *order* of the composition of infinitesimal motions does not matter—everything commutes—this is not true of the "finite" motions of a geometry. He thus sought to capture this "noncommutativity" at the infinitesimal level. The manner in which he did this is obscure in an interesting way because Killing needed to invoke the traditional views on infinitesimal motions while, in effect, denying their adequacy.

As was customary in the nineteenth century, Killing reasoned in terms of a basis, in this case a basis $dx = u_i\,dt$, $i = 1, 2, \ldots, m$, for the infinitesimal motions of the space form. Given two such motions, $dx = u_i\,dt$ and $dx = u_k\,dt$, according to the traditional interpretation, a point x is moved to $x + dx$ by the resultant composite motion, where $dx = (u_i + u_k)\,dt$. Although Killing did not mention this explicitly, he probably observed that if M_i and

M_k denote the corresponding finite motions, it can happen that $M_i M_k x \neq M_k M_i x$ and thus $(M_k M_i)(M_i M_k)^{-1}$ does not leave x fixed. Presumably this is why Killing decided that, at the infinitesimal level a nontrivial motion rather than the trivial $dx = 0$ implied by the traditional interpretation should correspond to the commutator motion $(M_k M_i)(M_i M_k)^{-1}$. We now consider how Killing obtained an infinitesimal analog of the commutator.

Consider first, the infinitesimal analog of $M_k M_i$. The motion $dx = u_i d\sigma$ sends x into $x + dx$, where, coordinatewise:

$$y_\rho = x_\rho + u_i^{(\rho)}(x)\, d\sigma, \qquad \rho = 1, \ldots, n. \tag{1}$$

The motion $dx = u_k dt$ then sends y into $z = y + dy$, where $dy = u_k^{(\rho)}(y)\, dt$. If, as was customary, $u_k^{(\rho)}(y) = u_k^{(\rho)}(x + u_i\, d\sigma)$ is expanded in a Taylor series and higher order terms neglected, the result is:

$$dy_\rho = \left(u_k^{(\rho)}(x) + \sum_{\nu=1}^{n} \frac{\partial u_k^{(\rho)}}{\partial x_v} u_i^{(v)}\, d\sigma \right) dt. \tag{2}$$

Thus the coordinates of the point $z = y + dy$ are:

$$z_\rho = x_\rho + u_i^{(\rho)}(x)\, d\sigma + u_k^{(\rho)}(x)\, dt + d\sigma \left(\sum_{\nu=1}^{n} \frac{\partial u_k^{(\rho)}}{\partial x_\nu} u_i^{(\nu)} \right) dt. \tag{3}$$

From the traditional standpoint, of course, the last term should be neglected since it is a second-order infinitesimal. The resultant equation (3) would then yield the usual interpretation of composition. Killing did not drop the term, although by separating $d\sigma$ and dt as he did, he seemed to wish to minimize the visibility of his unconventional procedure.

The infinitesimal analog of $M_i M_k x$ is obtained by the same procedure. Thus $x \rightarrow w$, where

$$w_\rho = x_\rho + u_i^{(\rho)}(x)\, d\sigma + u_k^{(\rho)}(x)\, dt + d\sigma \left(\sum_{\partial=1}^{n} \frac{\partial u_i^{(\rho)}}{\partial x_\nu} u_k^{(\nu)} \right) dt. \tag{4}$$

Now *by the traditional convention*, the inverse of $x \to x + dx$ is $x \to x - dx$. Accepting this convention, Killing obtained the infinitesimal motion $w \to z$ corresponding to $(M_k M_i)(M_i M_k)^{-1} w$ by subtracting (4) from (3) to obtain (with a convenient omission of the telltale dt):

$$dx_\rho = U_{ik}^{(\rho)} d\sigma, \qquad U_{ik}^{(\rho)} = \sum_{\nu=1}^{n} \left(u_i^{(\nu)} \frac{\partial u_k^{(\rho)}}{\partial x_\nu} - u_k^{(\nu)} \frac{\partial u_i^{(\rho)}}{\partial x_\nu} \right). \quad (5)$$

Since the motions $dx = u_i dt, i = 1, \ldots, m$, form a basis for all motions of the space form, this new motion (5) must be expressible in terms of them. That is, constants $a_{j,ik}$ must exist so that

$$U_{ik} = \sum_{j=1}^{m} a_{j,ik} u_j .. \qquad (6)$$

A total of m^2 motions $dx = U_{ik} d\sigma$ are produced in this manner, but at most m of them can be linearly independent. Realizing this, Killing sought relations among the U_{ik}. He singled out the following:

$$U_{ik} + U_{ki} = 0, \qquad U_{ii} = 0 \qquad (7)$$

$$\sum_{j=1}^{M} \left(a_{j.kl} U_{ij} + a_{j,li} U_{kj} + a_{j,lk} U_{lj} \right) = 0. \qquad (8)$$

Equations (6)–(8) look more familiar when translated into differential operator notation. Associate with the motions u_i and u_k the respective operators

$$X_i = \sum_{\rho=1}^{n} u_i^{(\rho)} \frac{\partial}{\partial x_\rho} \qquad \text{and} \qquad X_k = \sum_{\rho=1}^{n} u_k^{(\rho)} \frac{\partial}{\partial x_\rho}.$$

Then $[X_i X_k]$ corresponds to the motion U_{ik} and (6) corresponds to

$$[X_i X_k] = \sum_{j=1}^{m} a_{j,ik} X_j,$$

while (7) and (8) correspond respectively to:

$$[X_i X_k] + [X_k X_i] = 0, \qquad [X_i X_i] = 0$$

$$[X_i[X_k X_l]] + [X_k[X_l X_i]] + [X_l[X_i X_k]] = 0.$$

The infinitesimal motions of a space form thus constitute, in effect, a finite dimensional Lie algebra.

The problem of determining all space forms therefore involved three substantial subproblems. First there was the algebraic problem of determining all nonequivalent possibilities for the "structure constants" $a_{j,ik}$ or, in other words, the problem of determining all nonisomorphic Lie algebras. In the algebraic problem the u_i are treated as symbols, not specific functions of x. The analytic problem is to determine these functions by solving the partial differential equations given by equations (5) and (6). The third and final problem was geometrical: given the functions $u_i = u_i(x)$, and thus the infinitesimal motions of the space form, describe its geometrical characteristics.

The problem of providing an exhaustive determination of space forms was thus analogous to, albeit infinitely more difficult than, the problem resolved by Killing in his doctoral dissertation [1872]. In the present problem as well, the theory of elementary divisors enters in a natural way. In order to integrate the partial differential equations (5) and (6) it is desirable to choose constants $a_{j,ik}$ so that most are zero. One way to accomplish this is to begin by choosing a basis as follows. Choose, using the operator notation, X_1 arbitrarily (at least provisionally). Then choose the remaining X_i so that the linear transformation $X \rightarrow [X_1 X]$ is in its Weierstrass-Jordan canonical form. By this route Killing commenced to develop what was to become the theory of the structure of Lie algebras and especially semisimple Lie algebras.

Killing's theory of space forms as just described was published in 1884 as a *Programmschrift* [Killing, 1884], i.e. a scholarly essay appended, as was the custom, to the schedule of courses to be offered at the Lyceum during the forthcoming semester. It was consequently not widely read, if read at all. But Killing fortunately sent a copy to Klein, who, being a friend of Sophus Lie, suggested a possible connection between Killing's space forms and Lie's theory of continuous transformation groups, a theory he had been developing since 1874, although it was still (in 1884) not well known in Germany since Lie published most of his work in an inaccessible Norwegian journal that he edited. Killing finally obtained copies of these publications on loan from Lie for a relatively short time and he never really digested their contents. He read through them primarily to see where he had been anticipated by Lie, and he discovered that his own methods were not employed by Lie. Through his contact with Lie and Lie's assistant, Friedrich Engel, however, Killing's research received a needed focus. Lie published a paper [1885] on the application of transformation groups to differential equations and stressed the importance, in the Galois type theory he envisioned, of being able to specify all simple groups. Lie's methods were not suited to achieving such a specification, and Killing felt that his own methods might succeed here. And he was correct! For further details of this "success story," see [Hawkins 1982].

Postscript.   The above presentation is based upon a more extensive and carefully documented study [Hawkins, 1980], which, in particular, deals with a matter passed over here in silence: the apparent similarities between Killing's theory of space forms and Klein's *Erlanger Programm* of 1872. A careful examination of Klein's attitude towards mathematics and its relation to science and intuition suggests that the similarities are superficial and belie a more fundamental lack of affinity between Klein's *Erlanger Programm* and Killing's *Braunsberger Programm* [1884]. Although Klein did characterize the study of geometry as the study of transformation groups acting on manifolds, he did so in the hope of bringing unity to the seemingly disparate geometrical theories then proliferating and with no intention of calling for an exhaustive determination of all such groups in order to establish the foundations of non-Euclidean geometry. Geometry, and mathematics in general, was never to be pursued on a level or in a manner incompatible with intuition, experience and the needs for science. As the following passage from Klein's lectures on non-Euclidean geometry illustrates, he was totally unsympathetic to the attitude toward mathematics fostered by Weierstrass and his colleagues at Berlin:

With what should the mathematician concern himself? Some say: certainly intuition is of no value whatsoever; I therefore restrict myself to the pure forms generated within myself, unhampered by reality. That is the password in some places in Berlin. By contrast, in Göttingen the connection of pure mathematics with spatial intuition and applied problems has always been maintained and the true foundations of mathematical research recognized in a suitable union of theory and practice. [Klein, 1893, Vol. II, 361].

Despite the exaggeration, Klein's words captured the spirit in which mathematics was pursued in Berlin. The contrasts he drew between Berlin and Göttingen apply in particular to Killing's and his own approach to geometry.

REFERENCES

Biermann, K-R., 1973, *Die Mathematik und ihrer Dozenten an der Berliner Universität* 1810-1920, Akademie-Verlag, Berlin.

Hawkins, T., 1975, "Cauchy and the Spectral Theory of Matrices," *Historia Mathematica* 2, 1–29.

_____, 1977a, "Weierstrass and the Theory of Matrices," *Archive for History of Exact Sciences* 17, 119–163.

_____, 1977b, "Another Look at Cayley and the Theory of Matrices," *Archives Internationales d'Histoire des Sciences* 26, 82–112.

_____, 1980, "Non-Euclidean Geometry and Weierstrassian Mathematics: The Background to Killing's Work on Lie Algebras," *Historia Mathematica* 7, 289–342.

_____, 1982, "Wilhelm Killing and the Structure of Lie Algebras," *Archive for History of Exact Sciences* 26, 127–192.

_____, 1983, "Non-Euclidean Geometry and Weierstrassian Mathematics: The Background to Killing's Work on Lie Algebras," *Annals of the New York Academy of Sciences* 412, 73–83.

Helmholtz, H., 1868, "Ueber die Thatsachen, die der Geometrie zum Grunde liegen," *Nachrichten von der Gesellschaft der Wisenschaften zu Gottingen* Nr. 9. Reprinted on pp. 618–639 of [Helmholtz, 1883].

_____, 1883, *Wissenschaftliche Abhandlungen*, Vol. 2, Leipzig.

Killing, W., 1872, *Der Flächenbüschel zweiter Ordnung*, Inauguraldissertation Berlin.

_____, 1880, *Grundbegriffe und Grundsätze der Geometrie*, Programm des Gymnasiums zu Brilon.

_____, 1884, *Erweiterung des Raumbegriffes*, Programm Lyceum Hosiamum Braunsberg.

_____, 1885, *Die Nichteuklidischen Raumformen in analytischer Behandlung*, Leipzig.

_____, 1888a, 1888b, 1889, 1890, "Die Zusammensetzung der stetigen endlichen Transformationsgruppen," *Mathematische Annalen* 31, 252–290, 33, 1–48, 34, 57–122, 36, 161–189.

Klein, F., 1893, *Nicht-Euklidische Geometrie. Vorlesung, gehalten während 1889–1890*, Fr. Schilling (ed.), Göttingen (lithographed second printing).

Kronecker, L., 1874, "Ueber Schaaren von quadratischen und bilinearen Formen," *Monatsberichte der K. Akademie der Wissenschaften zu Berlin*, 349–413. Reprinted on pp. 349–413 of [Kronecker, 1895].

_____, 1895, *Mathematische Werke*, Vol. 1, Leipzig.

Lie, S. 1885, "Allgemeine Untersuchungen ueber Differential-gleichungen, die eine kontinuierliche, endliche Gruppe gestatten," *Mathematische Annalen* 25, 71–151. Reprinted on pp. 139–223 of [Lie, 1927].

_____, 1927, *Gesammelte Mathematische Abhandlungen*, vol. 6, Leipzig and Oslo.

Riemann, B. 1868, "Ueber die Hypothesen, welche der Geometrie zu Grunde liegen," *Abhandlungen der K. Gesellschaft der Wissenschaften zu Göttingen* 13. Reprinted on pp. 272–287 of [Riemann, 1892].

_____, 1892, *Gesammelte Mathematische Werke*, Leipzig.

Weierstrass, K., 1868, "Zur Theorie der quadratischen und bilinearen Formen," *Monatsberichte der K. Akademie der Wissenschaften zu Berlin*, 311–388. Reproduced with minor alterations on pp. 19–44 of [Weierstrass, 1895].

_____, 1895, *Mathematische Werke*, vol. 2, Berlin.

THE DISCOVERY OF
NON-EUCLIDEAN GEOMETRY

Jeremy Gray

INTRODUCTION

Few events in the history of mathematics compare with the discovery of non-Euclidean geometry. Throughout the 18th century anxiety about the foundations of Euclidean geometry had increased —the 'partie honteuse' of mathematics, as C. F. Gauss was to call it. In the first two decades of the 19th century several mathematicians came to believe that the formerly unthinkable might be true: a geometry different from Euclid's might, after all, be mathematically and physically possible. The published discoveries of non-Euclidean geometry by N. I. Lobachevskii [1829] and J. Bolyai [1831] nonetheless failed to win many converts for a generation. Only in the 1850s and 1860s did the mathematical community start to accept the new geometry, and by 1901 D. Hilbert could proclaim it as the most important discovery in geometry in the whole of the 19th century. In the 20th century non-Euclidean geometry has been successfully treated in various contexts: ergodic theory, special relativity, and topology of manifolds, to mention only three.

Such a long and complicated story invites many questions and has been written about from many points of view. To make sense of the curious history of non-Euclidean geometry we must consider, first, what geometry meant to the various protagonists; then there is the vexed question of the delayed acceptance of Lobachevskii and Bolyai's work; and, finally, we must ask why so many distinguished mathematicians, such as Legendre and, I would contend, Gauss, missed or misunderstood the new ideas.

Research into the foundations of geometry went through three overlapping phases. In the first phase, which culminated with Legendre, the basic concepts were taken as given (line, angle, plane, and so forth) and the style of reasoning was classical Euclidean mathematics. In the second phase, most ably represented by Lobachevskii and Bolyai, geometry was almost replaced by trigonometry which, it was obscurely seen, is somehow more basic. In the final phase the ideas of differential geometry were gradually seen to underlie the Euclidean concepts and, with differential geometry as a basis, the nature of both Euclidean and non-Euclidean geometry was elucidated. The physical importance of geometry was to the fore all the time; geometrical properties were assessed according to their physical significance or plausibility. The investigations, although plainly about assumptions, were about the assumptions needed to study space and were not conducted merely in a formal sense. In view of the two-fold reformulation of geometrical methods, I believe it is now easy to see why the journey from Saccheri [1733] to Beltrami [1868] took so long, and why the unexpected outcome of the story was a revolution in our ideas of geometry, in which Euclid's system came to be replaced with differential geometry on the one hand and truly rigorous axiomatic systems on the other.

1. EARLY INVESTIGATIONS OF THE PARALLEL POSTULATE

The fifth postulate of the *Elements* was Euclid's attempt to resolve a debate about parallel lines that was current in Aristotle's time. The postulate, in Heath's translation [Euclid's *Elements*, 1956, 202, postulate 5] asserts:

> That, if a straight line falling on two straight lines make the interior angles on the same side less than two right angles, the straight lines, if produced

indefinitely, meet on that side on which are the angles less than the two right angles.

The other assumptions made by Euclid are already strong enough on their own to show that at *least* one line can be drawn through a point P not lying on a line l which does not meet l (i.e., is parallel to it). The parallel postulate just quoted implies, with Euclid's other assumptions, that the line parallel to a line l through a point P not lying on l is unique [*Elements*, I, Prop. 31]. The postulate is very useful for deriving metrical propositions. With it Euclid could prove, for example, that the angle sum of any triangle is 180°, and that opposite sides of a parallelogram are equal [*Elements*, I, Prop. 33]. It follows from this result that a curve everywhere equidistant from a straight line is itself a straight line. Conversely, any of these results, taken with Euclid's other assumptions, imply the parallel postulate, and much of the early study of the foundations of geometry is a discovery and debate of these and related equivalences. Most frequently encountered is the assumption that the curve equidistant from a straight line is itself straight (it is found in such diverse places as Ibn al-Haytham (965–1039) and C. S. Clavio (1537–1612), an early editor of Euclid's *Elements*), but J. Wallis' (1616–1703) assumption, made on July 11, 1663 in his lecture at Oxford, that similar, noncongruent figures exist is another assumption of this kind. The common feature of all these assumptions is that they express a property of rectilinear figures which their authors believed to be true of the real world. Indeed, so strong was this belief that in some cases these assumptions were made only tacitly. But whether explicit or not, the end result of such arguments was to replace the parallel postulate with another assumption of equivalent effect, although sometimes of greater intuitive plausibility. For this reason, these researches can be considered scientific, for they represent alternative mathematical models of the physical world about which certain things are apparently known. At stake was the necessity for this or that assumption, given the other mathematical or physical assumptions already made. The hope, persistently frustrated, was that Euclid's assumptions— without the parallel postulate—would be enough to generate the postulate and its pleasing consequences. The outcome of these attempts was always the need for another starting point, although in some cases this may have escaped the notice of the authors of these works. The early authors never convinced all their readers, and at no time from 300 B.C. to A.D. 1850 was the problem of parallels widely considered to have been solved.

A good case in point is Gerolamo Saccheri (1667–1733), who is one of the last and greatest investigators in this tradition. In his *Euclides ab omni naevo vindicatus* (Euclid freed from every flaw, 1733) he attempted an ambitious defense of the parallel postulate. Starting from the observation that the postulate is equivalent to the assumption that the angles in a quadrilateral add up to 360°, he proposed to investigate three geometries that shared Euclid's assumptions except for the parallel postulate, which was replaced by one of the following assumptions about a quadrilateral:

(1) its angle sum exceeds 360°; or
(2) its angle sum is equal to 360° (Euclidean geometry again); or
(3) its angle sum is less than 360°.

Saccheri aimed to show by *reductio ad absurdum* that the first and third of these each "destroys" itself, as he put it, leaving the second, Euclid's geometry, true. He succeeded in the first case[1] and

---

[1] Since it is well known that spherical geometry exemplifies the condition that the angle sum of a quadrilateral exceeds 360°, this result of Saccheri's calls for some explanation. However, spherical geometry differs from Euclid's in two respects: it has no parallel lines and a line segment cannot be indefinitely doubled. Saccheri's first assumption represents an attempt to establish a geometry different from Euclid's only in the sense that it has no parallel lines; thus, spherical geometry notwithstanding, his proof can be valid. I should also point out that Saccheri's argument in this connection is ingenious. The contradiction in the geometry based on the assumption that the angle sum of a quadrilateral exceeds 360° is that in it the parallel postulate can be deduced as a theorem. Saccheri showed that, based on his first assumption, every quadrilateral has an angle sum of more that 360°; but the parallel postulate implies the angle sum is invariably 360°. Thus his contradiction follows. His proof in fact makes considerable use of the fact that the other assumptions of Euclid permit the arbitrary duplication of length.

A geometry on the sphere is sometimes called doubly elliptic geometry (the term is due to Klein [1892]) or singly elliptic if opposite points on the sphere are identified. Spherical geometry is not an absolute geometry, since it violates the axiom that two points determine a line.

then set to work on the third case, where he established the following results. Given a line *l* and a point *P* not on *l*, the lines through *P* divide into three families:

(a) an infinite family of lines meeting *l*;
(b) an infinite family of lines which never meet *l*;
(c) two lines which are asymptotic to *l* and separate the first two families.

Each line in the second family has a unique common perpendicular with *l* and diverges infinitely from *l* in both directions. Although these conclusions appeared to be unnatural, Saccheri knew they did not amount to a logical contradiction. To arrive at such a contradiction, he considered the behavior of a line, *m*, that passes through *P* and does not meet *l*, as it is rotated about *P*. He showed that there is no last position for *m*, but that the line can always be lowered a little further without leaving the second family. As it descends, its common perpendicular with *l* moves away from *P*, and Saccheri asked what happens in the limit. He persuaded himself, illegitimately, that, in the limit, *m* coincides with one of the two asymptotic lines—call it *k*—and that the common perpendicular of *m* with *l* now passes through the point of intersection of *m = k* and *l*, which he admitted is at infinity. Because the distinct lines *l* and *k* cannot have a common perpendicular at a common point, Saccheri proclaimed that the third hypothesis destroyed itself, and so Euclidean geometry was the only logical possibility. While no one seems to have fallen for Saccheri's mistake, it is worth noting the language he used in describing the contradiction as "repugnant to the nature of the straight line...." In this way Saccheri appealed to a body of beliefs about lines; it was somehow known that lines can't do certain things.

We see a similar process at work in Saccheri's most distinguished immediate successor, the Swiss Johann Heinrich Lambert (1728–1777). Lambert is most interesting for the ways in which he went beyond Saccheri, whose work he knew at least indirectly, because his friend A. G. Kaestner had persuaded him in 1763 to read G. S. Klügel's refutation of Saccheri's work and about 30 other accounts. Like Saccheri, Lambert discussed three geometries, each determined by the angle sum of quadrilaterals, and like Saccheri he showed that one could not logically assume the angle sum was greater than 2π. Since he had already ruled out the possibility of a geometry in which two straight lines enclose an area, he did not stop to investigate which other assumptions of Euclid's must be abandoned for spherical geometry to be possible. On turning to the geometry in which the angle sum of a quadrilateral is less than 2π, he found that the area of a triangle with angles α, β, and γ is proportional to $\pi - (\alpha + \beta + \gamma)$; noticing that a spherical triangle with angles α', β', γ', has an area of $R^2(\alpha' + \beta' + \gamma' - \pi)$, he said that he "could almost conclude that the new geometry would be true on a sphere of imaginary radius" (Lambert, in [Engel and Stäckel, 1895, §82]). However, he did not explain what such a surface might be, and his (unstated) argument presumably consisted only of replacing R by iR in the formula for the area, although he did not make even that substitution explicitly.

From his result about area, Lambert was able to deduce that in the new geometry similar figures must also be congruent. Accordingly there is, up to congruence, only one equilateral triangle with a given angle $\alpha < \pi/3$, and so there is a 1-1 correspondence between lengths and angles.

From this he deduced the remarkable conclusion that in the new geometry there is an absolute measure of length (as there is for angle in Euclid's). In other words, the measurement of length does not depend on the choice of an arbitrary length as the unit—there is a canonical choice on which all observers can agree without fear of making a change of scale. For a while Lambert thought this unexpected property of length gave him the contradiction he was seeking, for it was a property of quantity expressly denied by his philosophical mentor C. Wolff, a follower of Leibniz. But eventually Lambert decided to reject this philosophical *a priori* reasoning and to retain the mathematical conclusion without seeing it as a contradiction. In this way Lambert's work was robbed of a definite conclusion and, perhaps for that reason, he never published it—it was published posthumously in 1786 by his former colleague, Johann Bernoulli III. All Lambert provided by way of a conclusion was a list of suggestions for proving that the new geometry could not exist, which was evidently his hope. However, for explicitly exploring what the new geometry might look like and not giving a false refutation of that geometry, Lambert deserves to be considered as the first of the new investigators of the problem of parallels.

Not that the series of old–style investigators had come to an end. This series was to be consummated (but not concluded) by the distinguished Adrien-Marie Legendre (1752–1833). His very influential *Eléments de géométrie* was first published in 1794; it ran to at least 12 editions in his lifetime and 21 by 1876. His book was part of a back-to-Euclid movement which sought to attend to foundational questions, thus replacing an emphasis, derived from Cartesian philosophy, on not proving the obvious. Consequently Legendre had to deal with the problem of parallels, and in the various editions of his *Eléments* he dealt with it in different ways, not all of them of equal merit. Even the best of these attempts were not successful, of course, and all except one share one common feature: they are exercises in classical geometry. The concepts of line and angle were always assumed to be well understood, in the sense of scientific knowledge (as described in the Introduction). For Legendre, only the question of deducing the parallel postulate was troublesome. Indeed, Legendre would on occasion base his arguments on quite metaphysical, rather than mathematical, considerations, as for example in the 12th edition of the *Eléments*, [1823, p. 279] when he rested his arguments on a belief that it was "repugnant to the nature of the straight line that such a line extended indefinitely can be wholly contained within an angle," for such a line would not divide space into two equal pieces.

Although Legendre never changed his basic view of the problem, another mathematician who began his work from a strictly classical standpoint did eventually come to see the inadequacy of that position; this was Carl Friedrich Gauss (1777–1855). Whole books have been written on Gauss' views about non-Euclidean geometry ([Reichardt, 1976] can be recommended), and it is very difficult to summarize these views concisely. Gauss' earliest investigations, apparently done when he was 15 but not written down in permanent form until he was 55, explored the consequences of a geometry in which parallels are not unique. They are based on Lambert's work, which Gauss borrowed from the university library when he first went to the University of Göttingen in 1795 and again in 1797.

Gauss' views on the inadequacy of the classical approach were harsh: "Such treatments are as good as nothing," he wrote to W. Bolyai in 1799, and he doubted if the problem was capable of a logical solution. But if this is a recognition that a non-Euclidean geometry might be possible, it is a highly guarded one, and it was not accompanied by any vigorous investigation of the new terrain. Indeed Gauss never conducted an investigation of non-Euclidean geometry on anything like the scale with which he tackled, say, number theory, the theory of elliptic functions, statistics, astronomy, or, most interestingly for this discussion, differential geometry. (I shall return to this point in Section 3.) Instead, Gauss allowed himself to be drawn into correspondence with various nonprofessional mathematicians who were interested in the problem and with whom he was willing to share his unorthodox views.

The sudden profusion of people prepared to explore a non-Euclidean geometry is itself interesting and has never been satisfactorily explained. Wolfgang Bolyai (1775–1856), the father of Janos and a fellow student of Gauss' at Göttingen, is an example; F. K. Schweikart (1780–1859), a professor of jurisprudence, is another; and his nephew F. A. Taurinus (1794–1874) is a third. Indeed, there was a revival of interest in geometry underway at the time; in France, Monge's (1746–1818) teaching and Legendre's *Eléments* were its chief embodiments. Moreover, not only French mathematics was dominant in Europe; the contemporary success of Napoleon's armies was widely attributed in part to the training of military engineers at the new École Polytechnique. Although mathematics in Germany could not be nearly so coherently organized as it was in France—Germany was divided into several principalities, all of them disrupted by the war with Napoleon—yet, in Göttingen, A. G. Kaestner (1719–1800), who was the author of several books on the subject, had tried to keep geometry alive. So perhaps it is not surprising that the more disparate states of Germany, the small but growing mathematical communities in Central Europe, and the Russian school of mathematics took to problems in elementary geometry, while the scholars at the strong centralized schools in France—Legendre excepted—discussed the questions in quite perfunctory ways. Of course, most of the published discussions of these investigators were conservative in the sense that they sought to establish the logical necessity of Euclidean geometry; indeed, in 1816 Gauss lamented that scarcely a year went by without there being some flawed attempt to prove the parallel postulate. But some of these men did come to consider a new geometry in its own right and to reserve their judgement about its real validity. Whether this is just a good response to the way their calculations took them or whether one can see a hint of European radicalism, liberalism, or romanticism here, is surely impossible to say.

Most interesting of these new studies is that of F. K. Schweikart, which was communicated in 1818 via C. L. Gerling to Gauss. Schweikart investigated a geometry he called Astral geometry, in which parallels are not unique, and he showed that in such a geometry there is a maximum size for the altitude of a right angled triangle. He didn't regard this result as a logical contradiction; nor did Gauss, who replied in March 1819 to Schweikart that he had developed the new geometry so far that he could solve all the elementary problems within it, once a value for the maximum altitude was given. For example, the maximal area of a triangle is $\pi c^2/(\ln(1 + \sqrt{2}))^2$, where c denotes the maximum altitude. It is not clear what to make of Gauss' claim that he had developed the geometry this far. There is an abundance of notes made by Gauss and people who knew him (like Schumacher) that show he was floundering, certain only that the defenders of Euclid had failed to secure the crucial postulate. F. L. Wachter (1792–1817), a student of his who died in mysterious circumstances, made the beautiful discovery that in a three-dimensional, non-Euclidean geometry a surface called the horosphere (defined below in Section 2) carries a Euclidean geometry; his description was reprinted by P. Stäckel [Wachter, 1901]. Writing about this to H. W. M. Olbers in April 1817, Gauss said he inclined "more and more to the view that the necessity of our geometry cannot be proved." But two years later he wrote a note to himself, saying "In the theory of parallels we are still no further than Euclid. This is the partie honteuse of Mathematics, which sooner or later must be given a quite different form." Obviously Gauss did not believe that he had yet found the right solution to the problem. It would be entirely in keeping with everything else we know about Gauss' high standards to conclude

that he could see his way piecemeal through various elementary
arguments, but knew he did not possess the key idea around which
everything should be organized.

The classical style of argument, as I have called it, was just about
exhausted in the 1810s. By reasoning along Euclidean lines (pun
intended) investigators had been able to find several statements
equivalent to the parallel postulate and to derive several theorems
in a different geometry, one in which there are many parallels. But
no one had found a contradiction in this second geometry and, of
course, no number of new theorems could ensure that there was not
a contradiction waiting to be found. Since the axiomatic resolution
of the problem was blocked by the (generally tacit) assumption that
any acceptable non-Euclidean geometry had to be physically plausible, further progress could be made only by a better understanding of metrical geometry. Consequently, the dramatic change in
ideas about projective geometry brought about by the work of J. V.
Poncelet and A. F. Möbius was entirely irrelevant to the debate
about Euclidean geometry. Instead, the crucial developments were
born out of a prolonged period of obscurantism, which shall now
be described.

2. TRIGONOMETRIC STUDIES

There is a long, and relatively neglected, tradition in geometry in
which trigonometrical instead of Euclidean arguments are used.
Moreover, since trigonometry was largely developed for astronomical purposes, the theorems in spherical trigonometry have a long
history. The modern systematized theory of the trigonometric
functions is due to L. Euler (1707–1783), who was the first to show
(in 1741, but only published in 1748 in the *Introductio in analysin
infinitorum*) that

$$\sin x = \frac{e^{ix} - e^{-ix}}{2i}, \qquad \cos x = \frac{e^{ix} + e^{-ix}}{2}.$$

Inspired by this, Lambert defined in 1766 the hyperbolic trigonometric functions,

$$\sinh x = \frac{e^x - e^{-x}}{2}, \qquad \cosh x = \frac{e^x + e^{-x}}{2}$$

and showed how similar they were to the circular trigonometric sine and cosine. Although he used the new functions in his astronomical work, Lambert apparently never asked himself what kind of a triangle obeys the laws of hyperbolic spherical trigonometry. With hindsight this turns out to have been a near miss, we have to wait 60 years before anyone else was to consider the connection between the parallel postulate and spherical trigonometry. That man was F. A. Taurinus, a nephew of Schweikart and, like him, a lawyer.

Taurinus took one of the basic formulae of spherical trigonometry, which asserts that given a triangle on a sphere of radius R with sides a, b, c and angles A, B, C

$$\cos \frac{a}{R} = \cos \frac{b}{R} \cos \frac{c}{R} + \sin \frac{b}{R} \sin \frac{c}{R} \cos A,$$

and replaced R by iR. This gave him the fundamental formula of what he called his logarithmic spherical geometry:

$$\cosh \frac{c}{R} = \cosh \frac{b}{R} \cosh \frac{c}{R} - \sinh \frac{b}{R} \sinh \frac{c}{R} \cos A.$$

It is easy to see from this that if $a = b = c$, then $A < \pi/3$, so the formula must somehow be related to the many-parallel geometry. But the purely formal way in which the new formula had been derived obscures that connection, and Taurinus was unable to achieve sufficient clarity. He wrote two books on the subject; in the second he suggested that his hyperbolic formulae might describe the geometry on some surface or other, but he left the question there for other mathematicians to take up. He regarded the geometry as "so to speak inverse to spherical [geometry]" but gave strikingly poor reasons why it could not be a geometry of space. The honor of seeing that, on the contrary, the new geometry could be true belongs rightly and unreservedly to N. Lobachevskii (1793–1856) in Russia and J. Bolyai (1802–1860) in Hungary.

There is no doubt that Lobachevskii and Bolyai made their discoveries independently of each other and of Gauss,[2] despite the tenuous links connecting them to Göttingen (Lobachevskii's teacher at Kazan was Bartels, a friend of Gauss). That said, their discoveries and their arduous paths to them are strikingly similar. At the outset both believed they could prove the parallel postulate and, on discovering their mistakes, both were converted to the opposite belief (Bolyai in 1823, Lobachevskii in 1826); both were ready to publish before the decade was over. Lobachevskii published first, in the Kazan *Bulletin* for 1829, and Bolyai's work appeared as a 24-page appendix to his father's *Tentamen* [1831].

The crucial points common to both works are these:

1. The space studied is three dimensional.
2. Parallels in the same plane are not unique.
3. A surface called a horosphere is constructed as follows:
   a pencil of lines asymptotic in one direction to a given line is taken, and a surface is obtained which is perpendicular to all these lines. One such surface, called a horosphere, exists for each point in space. (The term is Lobachevskii's, Bolyai merely called it the surface F.)
4. Three lines in such a pencil are chosen, yielding a triangle on the horosphere, a triangle on a plane tangent to the horosphere, and, via a little further construction, a spherical triangle.
5. The triangle on the horosphere has an angle sum of π. (This is Wachter's earlier result.)
6. The formulae of spherical trigonometry are true without any assumption about parallels, and so are true in this new setting.

---

[2] This view, which expressly contradicts that of M. J. Kline [1972, p. 878], rests on the absence of non-Euclidean geometry in the Gauss-Bartels correspondence and the Gauss-Bolyai correspondence prior to 1832. Kline's conjecture that Bartels would have learnt from Gauss is refuted by the evidence analysed in [K. R. Biermann, "Die Briefe von Bartels an C. F. Gauss," *Naturwissenschaften, Technik und Medezin*, 1969, 10.1, 5–22].

7. Relations between the triangles allowed Bolyai and Lobachevskii to construct functional equations for the trigonometric functions that must describe the plane triangle.

8. These functions turn out to be the hyperbolic trigonometric functions; thus the new plane geometry is described by hyperbolic spherical trigonometry (i.e., Taurinus' formulae).

9. Consequently the elementary metrical properties of the geometry can be determined.

It is sometimes said that Bolyai's work differs from Lobachevskii's in placing a greater emphasis on theorems that are true independently of any assumptions about the parallel postulate, theorems in what Bolyai called absolute geometry. Although true, it is scarcely a major difference, whereas the similarities between their works are remarkable. Both regarded as an empirical matter the determination of the truth of either Euclidean or non-Euclidean geometry; Lobachevskii, in his first extensive study [1829, §151], went so far as to analyze observations on stellar parallax in an inconclusive attempt to resolve the matter. Moreover, strictly speaking both men based their confidence on nothing more than their original assumption about parallels; nowhere was this assumption shown to be free of contradictions. The fact that some trigonometric formulae are obtained at the end of their discussions is not logically enough, for the formulae might be lacking in geometric content. Lobachevskii was more aware of this problem than Bolyai, who published only this on the subject; in several subsequent works Lobachevskii came more and more to ponder the difficulty. He knew that the solution rested on obtaining a better understanding of the "straightness" of lines. His most extensive treatment is to be found in his *Neue Anfangsgründe der Geometrie* [1835] (the last chapter of which is more or less the same as his article in *Crelle's Journal* [1837]). Here Lobachevskii attempted to define surface and curve dimensionally by asserting that a surface separates space locally into two regions, and a curve separates a surface locally into two regions. But the best he could do for a straight line was, "A curve is called straight when it covers itself in all positions between two of its points" [Lobachevskii, 1837, 25]. So it may be said that he was trying to provide a better synthetic study of the elements of geometry than

had been provided hitherto, but it cannot be said that he succeeded.

The investigations of Lobachevskii and Bolyai did not at first
persuade anyone. The disappointing lack of response to Bolyai's
work can perhaps be put down to the obscure place where it was
published. Moreover, Gauss' reply to his 'old friend' Wolfgang
Bolyai when he had received the book was so sweeping that Janos,
convinced that Gauss was stealing his results, refused to publish
again. Gauss wrote, in an often quoted letter:

> If I began by saying that I cannot praise this you would be astonished, but I
> can do no other because to do so would be to praise myself, for the entire
> content of the work, the path your son has taken, and the results which he
> has obtained agree almost completely with my own meditations conducted in
> part for the last 30–35 years. [Gauss, 1900, 220].

Although he went on to say he was overjoyed that it was the son
of his old friend who had taken precedence over him in such
a manner, Gauss promptly made matters more painful by giving a
delightfully simple proof that in the new geometry the area of a
triangle with angles α, β and γ is proportional to $\pi - (\alpha + \beta + \gamma)$,
provided that this area is known to be bounded. Nor did Gauss do
much to spread the news abroad, beyond praising it to his friends.

It is, however, much harder to account for the failure of
Lobachevskii to catch the public eye. He published first in Russian
(in 1829 and 1835; German translations are in [Lobachevskii,
1899]), then in French in *Crelle's Journal für Mathematik* (1837)),
then a small German booklet, *Geometrische Untersuchungen zur
Theorie der Parallellinien* (1840), which has become the best known
of his works. He tried again in Russian and then again in French,
with his *Pangéométrie* (1855), the year before he died. The booklet
of 1840 was completely misunderstood by its reviewer in Gersdorf's
Repertorium (which cannot have helped), and the article of 1837
was frequently incomprehensible because it quoted results whose
proofs could be found only in the Russian article of 1835. This in
its turn provoked a fierce controversy in Russia that Gauss followed with interest (see his letter to Gerling of February 8, 1844,
on which this account is based [Gauss, 1900, 236]; Gauss eventually proposed Lobachevskii for membership in the Göttingen Sci-

entific Society). Lobachevskii's chief opponent was M. Ostrogradskii (1801–1861) who publicly ridiculed Lobachevskii's work; it is almost certain that Ostrogradskii's views held sway in the influential St. Petersburg Academy. Lobachevskii was, however, elected to the hereditary nobility in 1837 and was rector of Kazan University from 1827 to 1846, so he was a respected man—this was surely not because of his discovery of non-Euclidean geometry, but for his work in analysis and his energy as rector. In fact, his discovery was not to be widely respected until after his death, when in 1866 the energetic G. J. Hoüel published his French translation of the *Geometrische Untersuchungen*. By then the climate was receptive to his and Bolyai's ideas. In the next section I shall suggest why these changes took place.

3. DIFFERENTIAL GEOMETRIC FOUNDATIONS OF NON-EUCLIDEAN GEOMETRY

The debate about the nature and validity of non-Euclidean geometry could not be resolved before the introduction of a radical new idea. This is clear only with hindsight, because we know (now) where that idea was to come from: differential geometry. In this connection the decisive figure was Gauss. In 1827 he published his "Disquisitiones Generales Circa Superficies Curvas," in which he began to restructure differential geometry around the central idea that the (Gaussian) curvature of a surface is an intrinsic quantity, a result he had known since 1816. This means that its curvature at any point is determined solely by measurements which can be taken entirely on the surface and are independent of the space in which the surface may be embedded. The surfaces admitting the largest groups of isometries or, equivalently but in nineteenth century language, admitting the greatest degree of mobility for figures, are those of constant curvature.

The sphere has constant curvature, equal to $1/R^2$, where R is the radius of the sphere.

The cylinder and the plane have zero curvature, and when Gauss began to give his full attention to how the theory of curvature should properly be derived, he looked for a surface which had

constant negative curvature, "the opposite of the sphere," as he
called it. He found one example, the surface obtained by rotating
the tractrix[3] about its axis. However, Gauss did not at this stage
consider what the trigonometry on such a surface might be. Much
later, and apparently only after reading Lobachevskii's work, he
derived trigonometric formulae for a surface of constant curvature,
but he seems only to have had positive curvature in mind. Yet
again Gauss came close to putting the picture together, only to shy
away. Later D. Codazzi (1824–1873) showed that the trigonometry
on the pseudosphere was hyperbolic spherical trigonometry
[Codazzi, 1857], but he overlooked the connection with non-
Euclidean geometry. The pseudosphere is not the only surface of
constant negative curvature; indeed several surfaces of rotation
with that property had been exhibited by H. F. Minding
(1806–1885) ([Minding, 1839 and 1840]), who also showed that the
trigonometry of these surfaces was hyperbolic; his attention however, was focused firmly on their Gaussian curvature, which indeed
was a natural way to develop Gauss' ideas. These ideas were
brought together by B. Riemann (1826–1866) [1854] and E.
Beltrami (1835–1900), independently, [1868].

It is impossible to say with certainty what Riemann's attitude to
non-Euclidean geometry was. He did not mention it by name in his
famous lecture, "Ueber die Hypothesen welche die Geometrie zu
Grunde liegen," presented in 1854 to Gauss as his *Habilitations-
vortrag* (part of the examination that conferred the right to give
university lectures) but only published for the first time in late 1867
or 1868. Riemann spoke, however, of an obscurity, present since
Legendre's time, which lay at the heart of geometry, and he went
on to point out that there are three distinct 2-dimensional geometries of constant curvature; each characterized by the angle sum of
any triangle on that surface. From these remarks one may infer
that Riemann regarded non-Euclidean geometry as the geometry
on a surface of constant negative curvature, but it is equally clear

---

[3] The tractrix is the curve defined by the property that the segment of its tangent
TU lying between it and an axis *a* is of constant length, whence its name, since one
is supposed to walk along the axis *a* pulling a heavy weight *T* as one goes.

that this was far from his main point. He may have passed the subject over out of respect for Gauss' lifelong discretion on the subject, but it is more likely that he was interested in making a much more fundamental point: Riemann was concerned with establishing that any geometry on a manifold having dimension greater than one must be given an intrinsic formulation in terms of its curvature (suitably generalized). Even Gauss had considered only the geometry induced on a surface via its embedding in an ambient Euclidean three-space. Riemann is the first to regard intrinsic concepts as basic, thereby dethroning Euclidean geometry from its central position. Riemann's idea seems to have impressed Gauss greatly, to judge by his subsequent remarks to Weber (quoted in Dedekind's biography of Riemann available in Riemann's *Collected Works*, p. 549).

Beltrami's ideas were, on the other hand, directed precisely to sorting out the status of non-Euclidean geometry once and for all, as the title of his paper makes clear: "Saggio di interpretazione della geometria non euclidea" [Beltrami, 1868] (an essay on the interpretation of non-Euclidean geometry[4]). He began by observing that a geometry defined on a surface requires the free mobility of figures to establish congruences, and so the curvature of such surfaces must be constant. He thereupon investigated a surface of constant negative curvature and gave formulae for arc length and so forth for the geometry obtained by imposing constant negative curvature on the interior of a disc. This was easy for him to do, because the formulae closely resemble the case of spherical geometry. Then Beltrami showed that, with this new metric, straight lines appear as (Euclidean) chords of the disc, but in the non-Euclidean metric they have infinite length. Consequently, this is a many-parallel geometry—the differential geometric foundations of non-Euclidean geometry had at last been given! Now Beltrami derived Lobachevskii's trigonometric formulae (apparently he had not heard of Bolyai). This picture of the non-Euclidean plane is better than

---

[4]An English translation exists: J. C. Stillwell, Translation of Beltrami's "Essay on the Interpretation of non-Euclidean Geometry," Paper No. 4, Department of Mathematics, Monash University, Australia.

the one on the pseudosphere in the same way that the flat plane is better than the cylinder: it is simply connected and so contains no noncontractible closed curves. Beltrami knew that it is not accurate in all respects, any more than is a map of the globe one finds in an atlas (it is neither conformal nor area-preserving). However, because all lines appear in this model as lines, Klein was later able to hail Beltrami's picture as providing the right map for a projective model of non-Euclidean geometry [Klein, 1871]. It is nowadays known as the Beltrami-Klein model.

Two features of Beltrami's work deserve special comment. One is that he constructed a map of the non-Euclidean plane on a disc, without trying to construct a surface in Euclidean three-space which carried non-Euclidean geometry as the induced geometry. This was a wise move, because Hilbert was to show in 1901 that no smooth surface—i.e., one without singularities—can be embedded in three space in that way. (In the case of the pseudosphere, the rim is a curve of singular points beyond which the surface cannot be extended.) By exhibiting a map without any singularities Beltrami removed the last possible objection to non-Euclidean geometry— the anomalous nature of the known surfaces of constant negative curvature—for his disc is plainly as good a candidate for a picture of physical space as any Euclidean plane. The second feature of Beltrami's map is that it permits a simple proof of the independence of the parallel postulate from the remaining assumptions of Euclid; indeed, it can be directly verified that these assumptions are exemplified in the disc, while the parallel postulate is clearly false. A contradiction in non-Euclidean geometry would amount to showing that a certain construction in the disc was both possible and impossible. By carefully translating every step of that construction into Euclidean terms, this would imply that a corresponding Euclidean construction was also both possible and impossible, in short, that Euclidean geometry was itself self-contradictory. This argument, first made explicit by Bonola [1912, Appendix V], establishes the unexpected result that Euclidean and non-Euclidean geometries stand or fall together; either both are possible or neither. This conclusion was not anticipated prior to Riemann and Beltrami.

The work of Riemann and Beltrami provided an acceptable bedrock, and very few mathematicians thereafter were prepared to deny the validity of non-Euclidean geometry, although A. Cayley (1812–1895) continued to argue for the *a priori* Euclidean nature of physical space. Needless to say, philosophers could be found to disagree: G. Frege and J. Stallo, for example, took an unduly naive view of the geometrical possibilities of there being only one space. But until the publication of Beltrami's work mathematicians had been entitled to be sceptical, because of the obscurantism inherent in the trigonometry of Lobachevskii and Bolyai. As a Cambridge mathematician, Perronet Thompson, had argued in 1830 (quoted in [Halsted and Bonola, 1912, supplement p. 68]), if a triangle with circular arcs is used to prove or refute an argument about rectilinear triangles, one must demand "...to know, what property of straight lines has been laid down or established, which determines that what is not true in the case of other lines is true in theirs." In short, what is the crucial property of straight lines?

Before the invention of differential geometry, Thompson's question could not be answered; the various definitions of a straight line, from Euclid to Lobachevskii, did no more than point vaguely at the answer, leaving straightness as a primitive concept. Whereas, after Riemann one could proceed as follows: given a surface and a metric, one obtains the geodesics as the curves of a shortest length on that surface. If the surface is simply connected and has Gaussian curvature zero, then it is a plane and its geodesics are straight lines.[5] On surfaces of nonzero curvature the concept of a straight line is replaced by that of a geodesic, and in this way Perronet Thompson's question is answered.

But in the period from 1830–1868 the question of the nature of the achievements of Bolyai and Lobachevskii could and did hang in the air. For their trigonometric formulae to have a geometric

---

[5] Of course this is only true 'morally,' as one might say, and is not strictly true. A surface of zero curvature is locally isometric to a plane, but it may differ either by a bending or topologically, so its geodesics are only locally isometrically equivalent to straight lines. But these details do not alter the essential fact that only after the work of Riemann was it possible for mathematicians to say what a straight line is.

meaning it is necessary to be able to formulate all geometric concepts in analytic terms, and thus to give a Riemann-style treatment of geometry. The works of Bolyai and Lobachevskii inhabit a halfway house in which the primitive concepts of line, plane, and angle are, as in classical work, undefined, but their arguments are sustained by the formulae, which all but confer rigor on the proceedings.

4. CONCLUSIONS

It is Gauss' reformulation of the differential geometry of surfaces which eventually grew into a theory of geometric concepts capable of elucidating the problem of parallels, although Gauss did not conduct his studies with that aim in mind; whereas it is the trigonometrical work of Bolyai and Lobachevskii that made non-Euclidean geometry intuitively plausible and mathematically tractable. The unification of these different trends is the achievement of Beltrami, while Riemann's work provides an over-arching philosophy. In this connection it is interesting to note that many who accepted a two-dimensional non-Euclidean geometry continued to have doubts about the existence of a three-dimensional version (most notably J. Bolyai, for a while). Beltrami was not completely clear on this point in his 'Saggio,' which shows the power of three-dimensional Euclidean ideas coupled with the restriction— generally present in 19th century works—of differential geometry to the study of surfaces. However, while revising the 'Saggio' for publication Beltrami obtained Riemann's essay, "On the hypotheses which lie at the foundations of geometry." Realizing how these ideas could extend the tools of differential geometry to manifolds of any dimension, Beltrami [1870] was able to resolve the question of the existence of non-Euclidean geometries in spaces of any dimension (see [Milnor, 1982]).

The full significance of the discovery of non-Euclidean geometry cannot be discussed here, but one aspect requires further comment, namely its impact on the foundations of synthetic geometry. The discovery of non-Euclidean geometry inevitably casts doubt on the validity of Euclidean reasoning, as well as on the scientific truth of

Euclid's *Elements*. Simultaneously, in the 1830s and 1840s, projective geometers were conducting extensive, novel investigations of two- and three-dimensional real or complex projective spaces, often from a synthetic point of view. By the late 1880s it had become apparent that reasoning about lines in a geometry must not be based on tacit appeals to the nature of the line. The way in which this was accomplished was the development of an algebraic, even arithmetic theory coupled to an axiomatic approach. The axiomatic approach, which does not tell you what a line is but formalizes what can be done with it, is a distillation of the synthetic approach, and consistency of the axiomatic system was usually guaranteed by producing an algebraic model of it. This study was vigorously conducted in Italy (Peano, Pieri, Veronese) and Germany (Pasch, Hilbert); it rapidly became part of an extensive debate about rigor in mathematics, and so contributed to a change in our ideas concerning the nature of mathematics. It brought about a move away from the intuitive and towards the abstract and axiomatic. This move was concurrent with the writing of the main histories of the subject by F. Engel, P. Stäckel, and R. Bonola. Not unnaturally these historians tended to emphasize the axiomatic element in the story, and, as axioms permeated the mathematical curriculum, the history of non-Euclidean geometry was seen as a great historical example of a rival axiom system struggling to be born. Now that geometers have gained a better perspective on axiom systems it is easier to see the story as an attempt, in Morris Kline's words, to deal with "as fundamental a physical problem as there can be" [Kline, 1972, p. 881].

Non-Euclidean geometry is indeed a good example of an axiom system different from Euclid's which has its origins in questions about space. In that connection it is interesting how coolly astronomers regarded the question of whether space itself could be non-Euclidean. Gauss, Bessel, and Olbers were quite prepared for the answer to be yes, while admitting that the detection of decisive evidence might be beyond their abilities.

It is striking that the collapse of the oldest attempt to build a mathematical model of space around the assumptions of Euclid's *Elements* should lead directly to major changes in our views about the nature of space, geometry, and axiomatic reasoning.

Acknowledgements. I would like to thank H. J. M. Bos, U. Bottazzini, and E. Phillips for their helpful comments on an earlier draft. All translations are my own.

Annotated Bibliography

A subject as important as non-Euclidean geometry has naturally been treated in many ways, and I can only suggest some ways of exploring them here. The original historical work upon which almost all subsequent studies have relied was done by the distinguished German mathematicians F. Engel and P. Stäckel, and one can consult:

Engel, F. and P. Stäckel, 1895, *Die Theorie der Parallellinien von Euklid bis auf Gauss*, Teubner, Leipzig. (This book can be particularly recommended for its generous quotations from Wallis, Saccheri, Lambert, Gauss, and Taurinus. It is presently available as a Johnson Corporation reprint.)

Engel, F. and P. Stäckel, 1898, *Urkunden zur Geschichte der nichteuklidischen Geometrie*, Teubner, Leipzig. (This goes into the lives and works of the Bolyais' in great detail.)

Lobachevskii, N. I., 1899, *Zwei geometrische Abhandlungen*, Teubner, Leipzig. (A book prepared by F. Engel which also contains interesting commentaries.)

Stäckel, P., 1901, "Friedrich Ludwig Wachter." *Mathematische Annalen*, 54, 49–85.

Gauss, C. F., 1873, *Werke*, vol IV and 1900, *Werke*, vol VIII. (Engel and Stäckel helped edit Gauss' collected works, which contain his non-Euclidean research.)

Klein, F., 1871, "Über die so-genannte Nicht-Euklidische Geometrie," *Mathematische Annalen*, 4, = *Ges. Math. Abhandlungen*, I, 254–305.

Klein, F., 1892, *Vorlesungen über nicht-Euklidische Geometrie*, Teubner reprint 1927, Chelsea reprint (no date).

Legendre, A. M., 1823, *Eléments de géométrie*, 12th ed. Paris.

Saccheri, G., 1920, *Euclides ab omni naevo vindicatus*. (A reprint of the 1733 original as a parallel text, with translation and commentary by G. B. Halsted.)

Bonola, R., 1912, *Non-Euclidean Geometry, a Critical and Historical Study of its Development.* (First published as *La Geometria non-Euclidea* in 1906 and in translation in 1912 with the splendid addition of J. Bolyai's 'Appendix' and Lobachevskii's 'Geometrical Researches' [1840] translated and commented upon by J. B. Halsted, this history has also been the reference point for much work. A Dover edition of this book is presently available.)

Five recent English histories are available:

Gray, J. J., 1979a, "Non-Euclidean Geometry—a re-interpretation," *Historia Mathematica* 6, 236-258.

_____, 1979b, *Ideas of Space, Euclidean, Non-Euclidean, and Relativistic*, Oxford University Press.

Greenberg, M. J., 1978, *Euclidean and Non-Euclidean Geometries, Development and History*, Freeman.

Kline, M., 1972, *Mathematical Thought from Ancient to Modern Times*, Oxford University Press, Chapter 36.

Milnor, J. H., 1982, "Hyperbolic geometry: the first 150 years," *Bulletin of the American Mathematical Society*, (2) 6.1, 9–24. (This goes considerably beyond these treatments in keeping with modern interest in the subject.)

Works on the differential geometric foundations of non-Euclidean geometry referred to in Section 3 are listed below. Full references to the works of these authors may be found in [Bonola, 1912] and [Gray, 1979a, b].

Beltrami, E., 1868, "Saggio di interpretazione della geometria non-Euclidea," *Giornale di Matematiche* 6, 248–312. *Opere Matematiche*, I, 375–405, Zurich, ed. A. Speiser.

Codazzi, D., 1857, "Intorno alle superficie, le quali hanno costante il prodette de' due raggi di curvatura," *Annali di Scienze Matematiche e Fisiche* 8, 346–355.

Minding, F., 1839, "Wie sich entscheiden lässt, ob zwei gegebene krümme Flächen aufeinander abwickelbar sind oder nicht; nebst Bemerkungen über die Flächen von unveränderlichem Krümmengsmasse," *Journal für Mathematik* 19, 370–387.

——————, 1840, "Beiträge zur Theorie der kürzesten Linien auf krummen Flächen," *Journal für Mathematik* 20, 323–327.

Riemann, B., 1867, "Über die Hypothesen welche der Geometrie zugrunde liegen," *Abhandlungen der Königlichen Gesellschaft der Wissenschaften zu Göttingen* 13, reprinted in *Gesammelte mathematische Werke und wissenschaftlicher Nachlass*, R. Dedekind and H. Weber (eds.), 2nd edition, Leipzig, 1902, Dover reprint 1953, 272–287.

A modern exponent of the axiomatic treatment, who also is knowledgeable about the connection with theology, is Imre Toth; see, for example,

Toth, I., 1982, "Gott and Geometrie, Eine viktorianische Kontroverse," *U.R. Schriftenreihe der Universität Regensburg*, 7, 141–204.

The specifically English aspects are well explored in

Richards, J., 1979, "The Reception of a Mathematical Theory: non-Euclidean Geometry in England, 1868–1883," in B. Barnes and S. Shapin (eds.), *Natural Order: Historical Studies of Scientific Culture*, Sage Publications, Beverley Hills.

For an account of how the history of non-Euclidean geometry fits into the bigger picture of the development of 19th century mathematics see

Scholz, E., 1980, *Geschichte des Mannigfaltigkeitsbegriffs von Riemann bis Poincaré*. Birkhäuser, Boston and Basel.

Some idea of the immensity of the problem and the number of works devoted to it can be gleaned from the size of:

Sommerville, D. M. Y., 1911, *Bibliography of non-Euclidean geometry*, p. 403.

On Greek matters, one can start with
Toth, I., 1967, "Das Parallelproblem im Corpus Aristotelicum," *Archive for History of the Exact Sciences*, 3, 249–422.

On the fascinating Arab story the initial reference must be the admirable work of the Moscow School,
Youschkevitch, A. R., 1964, *Geschichte der Mathematik in Mittelatter*, Leipzig.

Early Western work has only recently been considered in
Maieru, L., 1982, "Il Quinto Postulato Euclideo da C. Clavio [1589] a G. Saccheri [1733]," *Archive for History of Exact Sciences*, 27.4, 297–334.

I have been unable to obtain a copy of Rosenfeld's biography of Lobachevskii, but I can recommend
Rosenfeld, B. A., 1973, "Lobachevskii," *Dictionary of Scientific Biography*, vol VIII, 428–435.

A recent biography of Gauss is
Kaufmann Bühler, W., 1981, *Gauss: A Biographical Study*, Springer-Verlag, Berlin, Heidelberg, New York.

A study of Gauss's non-Euclidean geometry is to be found in
Reichardt, H., 1976, *Gauss und die nicht-euklidische Geometrie*, Teubner, Leipzig.

The most illuminating study of Gauss' differential geometry is
Dombrowski, P., 1979, 150 *Years After Gauss' "Disquisitiones generales circa superficies curvas,"* Astérisque 62. (This also contains Gauss' original text and the English translation by A. Hiltebeitel and J. Morehead.)

Other authors and other approaches exist, and I apologize for their omission, pleading only that non-Euclidean literature, like the subject itself, diverges infinitely in all directions.

L. E. J. BROUWER'S COMING OF AGE AS A TOPOLOGIST

Dale M. Johnson

Abstract

This paper describes the early development of L. E. J. Brouwer (1881–1966) as a topologist during the crucial period from around the time of his doctoral thesis, *On the Foundations of Mathematics*, to the time of his discovery of new and revolutionary methods and results, that is, from about 1907 to the middle of 1910. It covers some of his very early work that motivated his interest in topological problems. It then treats his critical handling of A. Schoenflies' topology, which spurred him to investigate the field of topology directly. Most importantly, it examines his discovery of the concept of mapping degree, which became his principal tool in topological work, and some of his first uses of this concept. Mention is also made of the scope of his most significant later topological work.

1. INTRODUCTION: THE BACKGROUND TO BROUWER'S WORK IN TOPOLOGY

At the beginning of the twentieth century two mathematicians prepared the way for the magnificent development of topology in

61

our times: Henri Poincaré and L. E. J. Brouwer. Poincaré
(1854–1912) erected a combinatorial and algebraic framework for
topology in his homology theory and homotopy theory, which he
revealed in a major series of papers that appeared between 1895
and 1904. Brouwer (1881–1966) viewed topology in quite a different way from Poincaré. Operating within a framework of set theory
and point set theory, Brouwer pushed topology to new limits with
his theories of mappings, degree, and dimension. He produced his
major papers between 1909 and 1913.

The aim of this paper is to describe Brouwer's early development
as a topologist up to the time of his discovery of new and
revolutionary methods and results, that is, up to about the middle
of 1910. The main sources that I have used in preparing this paper
are the documents, letters, and papers of the Brouwer Archive;
Hans Freudenthal's edition of Brouwer's *Collected Works*, volume
2, with its excellent historical notes [1976]; and my papers on the
history of topological dimension theory [Johnson, 1979; 1981].[1]

To begin I should point out some aspects of the background to
Brouwer's topological work. During the last three decades of the
nineteenth century Georg Cantor (1845–1918) created point set
theory. This special part of set theory with both analytical and
geometrical aspects had many diverse applications.[2] Eventually the
theory developed into point set or set-theoretic topology. When
dealing with problems of analysis, Camille Jordan (1838–1922)
attempted to prove the Jordan curve theorem asserting that every
simple closed planar curve, the homeomorphic image of a circle,
divides the plane into two domains, an interior domain and an
exterior domain. He published his proof twice, first in 1887 and

[1] I am grateful to Professor Dr. Hans Freudenthal, Professor Roger Cooke, and
Dr. Walter P. van Stigt for critically reading a draft of this paper. I have benefited
from their comments in preparing the final paper.

For Brouwer's works I have referred to the *Collected Works* [1975; 1976]. These
two volumes conveniently contain nearly all of Brouwer's published works in
photographic reproduction and have extensive notes and bibliographies. The Brouwer
Archive is currently located at the Rijksuniversiteit Utrecht, The Netherlands.

[2] See [Cantor, 1932], [Dauben, 1979], and [Johnson, 1979].

again in 1893. His proof and related work strongly influenced the subsequent development of set-theoretic topology.[3] In producing his famous space-filling curve Giuseppe Peano (1858–1932) effectively called attention to some of the hazards to be encountered in the topology of curves [Peano, 1890]. Around the turn of the century Arthur Schoenflies (1853–1928) was the foremost exponent of Cantor's set theory and point set theory. Relying on the work of Cantor, Jordan, Peano, and others, Schoenflies proposed and partly developed a special program of set-theoretic topology or *analysis situs* (to use the term then current). In several publications, especially a series of three 'Contributions to the theory of point sets' ('Beiträge zur Theorie der Punktmengen') [Schoenflies, 1904; 1904a; 1906] and the second part of the *Report, The Development of the Theory of Point Manifolds* (*Bericht, Die Entwickelung der Lehre von den Punktmannigfaltigkeiten*) [1908], Schoenflies intended to characterize the topology of the plane. Central to his planar topology were the Jordan curve theorem and its converse. He was the first to state and prove a converse to the closed curve theorem. His converse asserted that a closed curve dividing the plane into two domains, such that each of its points is *accessible* from both domains, i.e., such that each of its points can be connected by simple paths or arcs with the points of the two domains, is homeomorphic to a circle.

When Brouwer was beginning his career as a mathematician, set-theoretic topology could not be regarded as an especially well-developed part of mathematics, in spite of Schoenflies' efforts. Controversy surrounded Cantor's general set theory because of the paradoxes or contradictions. Point set theory was widely applied in analysis and somewhat less widely applied in geometry, but it did not have the character of a unified theory. Schoenflies was struggling to shape the theory into a coherent and well-defined piece of topology. However, we shall soon see that the foundations of his

---

[3] Bolzano seems to have been the first to consider the Jordan curve theorem; cf. [Johnson, 1977]. For Jordan's proofs of the closed curve theorem cf. [Jordan, 1887, 587–594; 1893, 90–100]. For Brouwer's sharpened statement of the theorem see Section 4.

topology were very shaky. The time was right for a further thrust of development in topology. We now come to Brouwer.

2. BROUWER AT THE BEGINNING OF HIS MATHEMATICAL CAREER

Brouwer's doctoral thesis of 1907, *On the Foundations of Mathematics* (*Over de Grondslagen der Wiskunde*) [Brouwer, 1975, 11–101] can be said to mark the beginning of his mathematical career.[4] The work revealed the twin interests in mathematics that dominated his entire career: his fundamental concern with critically assessing the foundations of mathematics, which led to his creation of intuitionism and intuitionistic mathematics, and his deep interest in geometry and geometrical constructions, which led to his seminal work in topology. In the geometrical part of the thesis Brouwer [1975, 15–52, Chapter 1] dealt with foundational issues related to the continuum and various geometries. The theory of groups of transformations, largely derived from the work of Sophus Lie, was fundamental to his investigations. In particular, he took up the Riemann-Helmholtz-Lie space problem [Brouwer, 1975, 30–41], and this led him to consider the related problem of the possible elimination of assumptions of differentiability from Lie's theory of continuous groups, in other words, Hilbert's fifth problem. Brouwer [1975, 39–40; cf. 1976, 102–108] analyzed Hilbert's attempt to characterize planar geometry in a topological way by certain groups of transformations ('On the foundations of geometry' ('Ueber die Grundlagen der Geometrie') [Hilbert, 1903]).

After completing his thesis and attaining his doctorate Brouwer turned primarily to geometrical studies. His contributions to topology were the most important outcome of these studies. Dr. Walter P. van Stigt [1979] has put forward a very plausible explanation for the fact that Brouwer did not immediately develop his intuitionistic ideas and instead concentrated his efforts for some time in the

---

[4] Brouwer published a number of works before completing his thesis and these are of some interest [1975, 1–10; 1976, 1–89]. However, the thesis stands as the real beginning of his work in mathematics and his philosophy of mathematics.

fields of geometry and topology. Though passionately interested in the foundations of mathematics, he quickly found that his general philosophical and special intuitionistic ideas sparked controversy. D. J. Korteweg, Brouwer's *promotor* (thesis supervisor), was not pleased with the more philosophical aspects of the thesis; he demanded that several parts of the original draft be cut out for the final presentation.[5] Afterwards Korteweg urged Brouwer to concentrate on more 'respectable' mathematics, so that the young mathematician might increase his standing among his Dutch colleagues and further his academic career. Apparently taking his teacher's advice, he set out to solve some really hard problems of mathematics, in order to make a name for himself. He bided his time with regard to intuitionism until securing a more permanent academic position. As a result he produced a flood of papers on continuous group theory and topology over the next few years.

3. BROUWER'S WORK ON HILBERT'S FIFTH PROBLEM AND HIS CRITIQUE OF SCHOENFLIES' *ANALYSIS SITUS*

Starting from the geometrical research of his thesis, Brouwer mounted a full-scale attack on Hilbert's fifth problem, that of building up the Lie theory of finite continuous groups independently of conditions of differentiability. He presented his initial results on the problem at the Fourth International Congress of Mathematicians held in Rome, 6 to 11 April 1908. By attending this gathering he came into contact with some of the leading mathematicians of his day. He was especially impressed by the French. Poincaré's contribution on the future of mathematics was awe-inspiring. No doubt Jacques Hadamard (1865–1963) was impressive too. It is even possible that Brouwer became acquainted with this Frenchman at the Rome Congress. Later Hadamard played a crucial role in the development of his topological ideas.

---

[5] Cf. [Stigt, 1979a].

For Brouwer the fundamental problem of Lie continuous group theory was one of classification:

> to determine all finite continuous groups of an n-dimensional manifold [Brouwer, 1976, 109],[6]

while meeting the requirement of Hilbert's fifth problem to eliminate assumptions of differentiability. His methods of attack on the problem were direct, primarily topological. He stripped away the analytical support used by Lie. In his Rome lecture [Brouwer, 1976, 109–117] and two subsequent papers [Brouwer, 1976, 118–179] he presented his attempts to determine these groups acting on 1- and 2-dimensional manifolds. Though Brouwer achieved only very limited results, he did take the problem in a new, distinctly topological direction. For example, he started with a new definition of topological manifold. His broad attack on the problem of constructing Lie's group theory according to the requirements of Hilbert's fifth problem soon prompted his investigation of many related topological problems, such as the problem of invariance of dimension for manifolds.

For his investigations of Lie groups acting on 2-dimensional manifolds Brouwer relied on the topological results of Arthur Schoenflies. Around the time of the Rome Congress he was engaged in a thorough study of Schoenflies' planar *analysis situs*, as contained in the series of articles entitled 'Contributions to the theory of point sets' [Schoenflies, 1904; 1904a; 1906] and the second part of the *Report* [Schoenflies, 1908]. At Rome Brouwer was proud to say that his group-theoretic investigations were probably the first instance of an application of the results of Schoenflies. Later during 1908 he presented a full review of topology in the style of Schoenflies in a lecture on 'Planar curves and planar domains' ('Vlakke krommen en vlakke gebieden'), which he delivered to the Dutch Mathematical Society in October.[7] In the lecture Brouwer critically examined the main themes of Schoenfliesian topology. Jordan's theorem on the separation of the plane by a

---

[6] When necessary I have translated the various quoted passages, including the one now given. My translations can be compared with the originals in the given references.

[7] Cf. [Johnson, 1981, 128].

closed curve and Schoenflies' converse to the Jordan theorem were of central importance. More generally, he looked at Schoenflies' concept of closed curve, defined as a bounded closed point set that divides the plane into two domains such that it is the common boundary of the two domains. He gave considerable attention to the topological invariance of closed curve and related matters. He also mentioned the topological invariance of dimension. Brouwer's review of the problem situation in planar topology was not overly critical of Schoenflies, though he did criticize Schoenflies' proof of the invariance of closed curve. During 1908 Brouwer was generally sympathetic to the ideas and methods of Schoenflies. His works of this time were a prelude to his more critical and deeper investigation of the field.

Shortly after delivering the lecture in October 1908 Brouwer discovered some very serious flaws in the topological work of Schoenflies. In a letter to Hilbert of May 14, 1909 he explained how he came upon the radical inadequacies of Schoenfliesian topology.

> During the previous winter when I had already completed my second communication on finite continuous groups for submission to the editorial board of *Mathematische Annalen*, I discovered all of a sudden that the Schoenfliesian investigations concerning the *analysis situs* of the plane, on which I had relied in the fullest way, could not be taken as correct in all parts, so that my group-theoretic results also became doubtful.
>
> In order to achieve clarity again it was first necessary to examine thoroughly the corresponding Schoenfliesian theory and to ascertain exactly which of his results could be completely trusted. In this way the enclosed work arose, which interferes in several parts of the Schoenfliesian theory and as it were reorganizes some of them. [Brouwer, 1976, 371]

The enclosure was the first submitted version of the famous paper 'On *analysis situs*' ('Zur Analysis Situs') [Brouwer, 1976, 352–370]. It displayed a set of unexpected and quite devastating counterexamples, including the first indecomposable continua and a curve that divides the plane into more than two domains but is the exact boundary of each of these domains. These counterexamples virtually demolished Schoenflies' planar topology. Brouwer sent a copy of his paper to Schoenflies as well as to Hilbert. Correspondences among the three men resulted. Schoenflies found it particularly difficult to comprehend Brouwer's deeper insights into the topology of the plane. In a letter to his *promotor* Korteweg of June 18, 1909 Brouwer wrote of the impact of his criticism on Schoenflies:

> At last some fish has taken the bait! ... I am so glad finally to receive something more than just a polite postcard about my work. ... Schoenflies has gone into my paper in considerable detail, but I had to put the thumbscrews on rather hard.[8]

With the paper, 'On *analysis situs*' Brouwer attained a new mastery of the field of topology—a mastery quite beyond the powers of Schoenflies.

Brouwer's example of an indecomposable continuum, i.e., a continuum that cannot be divided into two proper subcontinua, is derived from Figure 1 [Brouwer, 1976, 354]. The figure illustrates a stage in the construction of the indecomposable continuum. The construction proceeds by alternately putting a black (darker shaded) strip or 'channel' into the large rectangle and extending the red (lighter shaded) channel in the large rectangle, such that the channels come closer and closer together according to well-defined proportions and eventually, i.e., at the limit, fill the large rectangle. The limit-boundary of the black and red domains constitutes the indecomposable continuum K. Thus after sufficient continuation of the process of construction the red domain comes arbitrarily close to every point of the boundary of a black domain and the black domains come arbitrarily close to every point of the boundary of the red domain. The red and black domains are completely separated from one another by the complete boundary K of the red domain. Indeed the complementary set of K contains only two domains: (1) the red domain and (2) the union of the black domains and the domain outside the large rectangle. K is a closed curve in Schoenflies' sense; it is the common boundary of exactly two domains in the plane. Moreover, K is indecomposable. Any two points of K, for example, Q_1 and Q_2, cut the curve into arcs, at least one of which is improper. Actually in the case of the arcs determined by Q_1 and Q_2, both are improper; each arc contains the entire closed curve K.

---

[8] Letter quoted in [Stigt, 1979, 364–365].

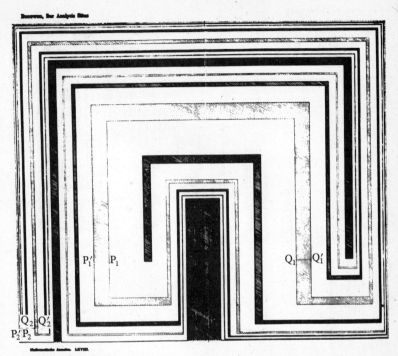

FIGURE 1. From Brouwer's paper 'On *analysis situs*' ('Zur Analysis Situs'). This is a monochrome photograph of his color Figure 2. The lighter shaded connected domain was red in the original, while the darker shaded domains were black.

Brouwer's example of a nowhere dense connected set that divides the plane into more than two domains and is the common boundary of all of them is derived from Figure 2 [Brouwer, 1976, 355]. For this counterintuitive example consider only the upper half of the main figure. The process for constructing the point set is similar to that for constructing the indecomposable continuum. The red (lighter shaded) and black (darker shaded) domains eventually cover the upper, nearly rectangular part of the figure, yielding a limiting boundary curve. Imagine that the upper part is deformed, so that the points F and K coincide and the points E and L coincide, while the rest of the geometrical structure of the upper

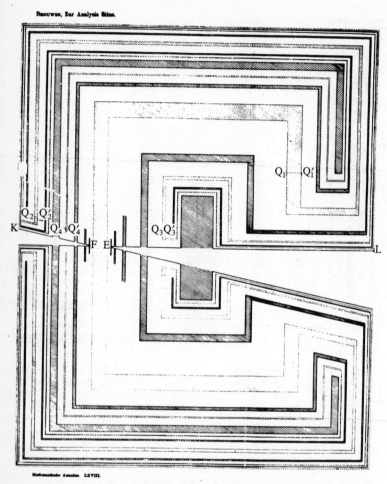

FIGURE 2. From Brouwer's paper 'On *analysis situs*' ('Zur Analysis Situs'). This is a monochrome photograph of his color Figure 3. The lighter shaded domains were red in the original, while the darker shaded domains were black.

part of the figure is preserved. In this way the plane is divided into three domains, a red domain, which is extended to include the rest of the plane outside the upper part of the figure, and two black domains, all having the same boundary. Moreover, it is possible to generalize this example, so that arbitrarily many domains, even infinitely many domains, have a common boundary.

In the last pages of 'On *analysis situs*' Brouwer subjected Schoenflies' topological work to a detailed critical analysis based on his counterintuitive examples. He pointed out several theorems of Schoenflies that were false, others that had defective proofs, and still others that were doubtful. His catalogue of errors and deficiencies in Schoenflies' work was devastating.

His spectacular example of a curve dividing the plane into three domains cast doubt on Schoenflies' closed curve theorem, which asserts that the topological image of a Schoenfliesian closed curve is again a closed curve. Might not a Schoenfliesian closed curve be transformed into a curve separating the plane into three or more domains? Brouwer took a special interest in this theorem. Eventually he devised a deep proof for it, which was published in 1912.

The results of 'On *analysis situs*' clearly showed the need for a new and much more rigorous approach in topology. This need strongly motivated Brouwer to reshape the field of topology.

Closely related to 'On *analysis situs*' were Brouwer's first and second papers [1976, 341–351; 384–395] in the series 'On the structure of perfect sets of points.' The first was communicated on March 26, 1910, the second, on April 28, 1911, to the Dutch Academy. Cantorian in their treatment of point set theory and topology, the papers effectively continued the critical treatment of Schoenfliesian results. In the first paper Brouwer analyzed the structure of the closed and perfect (and, hence, compact) sets in Euclidean *n*-space. He demonstrated the following important result.

Theorem. Each perfect set of pieces possesses the geometric type of order ζ.
[Brouwer, 1976, 344–346]

A *piece* is a maximal connected, closed (compact) set, i.e., a component. A *perfect set of pieces* is a closed set such that each piece is a limiting piece. The theorem means that a perfect set of such pieces or components can be put into one-one correspondence with the Cantor discontinuum (which in Brouwer's terminology has geometric type of order ζ); each piece corresponds to a point of the discontinuum with limiting relations preserved. Hausdorff reworked this result in the *Fundamentals of Set Theory* (*Grundzüge der Mengenlehre*) [1914, 322–324], and so it passed into the mathematical literature. A modern version of it reads as follows.

> The Cantor set is the only totally disconnected, perfect compact metric space (up to homeomorphism).   [Willard, 1970, 217]

In his first paper Brouwer also analyzed the Cantor discontinuum from a group-theoretic viewpoint, showing how the discontinuum gives rise to infinitely many topological groups of transformations [Brouwer, 1976, 346–350]. In this example we have the first nontrivial example of a topological group that is not a Lie group.

In the first paper Brouwer provided a proof of Schoenflies' generalization of the Cantor-Bendixson theorem. In the second paper he proved his well-known reduction theorem, which is an extremely broad generalization of the Cantor-Bendixson theorem [Brouwer, 1976, 384–385].

In these two papers Brouwer put the first fruits of his critical treatment of Schoenfliesian topology.

4: NEW DIRECTIONS IN BROUWER'S TOPOLOGICAL THINKING

By the middle of 1909 Brouwer was deeply engrossed in the problems that topology presented to the working mathematician. His work on the transformation groups of Lie together with some related investigations of transformations of surfaces and of vector fields (to be discussed below) had brought him to the very boundary of topology. His critical assessment of Schoenfliesian planar topology had warned him of some of the pitfalls to be avoided when entering this treacherous field. The occasion of his inaugural lecture as *privaat-docent*, i.e., unpaid lecturer, in the University of Amsterdam, *The Nature of Geometry* (*Het Wezen der Meetkunde*) [Brouwer, 1975, 112–120], gave him an opportunity to delineate the problem situation in topology and more generally to present his views on the foundations and philosophy of geometry. He delivered the lecture on October 12, 1909. Three days later he wrote a letter to Hilbert, which can be looked upon as a supplement to the lecture [Brouwer, 1976, 375–376]. Together the lecture and the letter outlined a program of research in topology—a program that he proceeded to carry out with great brilliance.

In the first part of his inaugural lecture Brouwer considered the nature of geometry from a very broad philosophical perspective, inquiring into the relation of geometry to experience and experimental science and to the rest of mathematics. He reviewed various philosophical positions concerning the nature of geometry that have been taken since Kant. He stressed the impact of the discovery of many new geometries during the nineteenth century on these philosophical positions. Upon inquiring into the possibility of an *a priori* basis for geometry, Brouwer put forward his fundamental intuitionistic view of the primary foundation for geometry.

> Must it be concluded that there is no *a priori* form of perception at all for the world of experience? There is, but only in so far as any experience is perceived as spatial or nonspatial *change*, whose intellectual abstraction is the *intuition of time* or *intuition of two-in-one*. From this intuition of time, independent of experience, all the mathematical systems, including spaces with their geometries, have been built up, and subsequently some of these mathematical systems are chosen to catalogue the various phenomena of experience. [Brouwer, 1975, 116]

Thus for Brouwer the intuition of time gave rise to the construction of the mathematical continuum and the various multidimensional spaces.

Brouwer put forward a definition of geometry, in order to differentiate the subject from other parts of mathematics.

> *Geometry is concerned with the properties of spaces of one or more dimensions. In particular, it investigates and classifies sets, transformations, and transformation groups in these spaces.* [Brouwer, 1975, 116][9]

He added significantly that the spaces to be considered in geometry should be built from one or more Cartesian simplexes with various types of connection. He meant that simplicial complexes and

---

[9]Italics in original.

polyhedra and perhaps more particularly manifolds should be
regarded as the primary substrate for geometries. He adopted this
view in his subsequent research in geometry and topology. He soon
proposed a new definition of *n*-dimensional manifold as a foundation for his topological investigations. Though he mentioned groups
of transformations in the definition of geometry given in the
lecture, as he delved more deeply into topology he moved away
from the 'Kleinian' transformational viewpoint. Transformation
groups became far less significant in his study of *n*-dimensional
topology.

In his inaugural lecture Brouwer devoted most of his attention to
surveying the field of *analysis situs* or topology. By the time of
giving the lecture he was able to mention some of his own conjectures and results. He identified the fundamental problem of *analysis situs* as the problem of giving

> a *classification for analysis situs of the sets of points in a space*.   [Brouwer,
> 1975, 118][10]

according to the group of *analysis situs*, the group of one-one
continuous (and onto) transformations.[11] This classification problem was central to Riemann's (and also Poincaré's) conception of
topology, though it was not the main problem treated in Brouwer's
subsequent work. However, he mentioned the following:

> An immediately related problem is, in how far spaces of different dimension
> are different for our group. Most probably this is always the case, but it
> seems extremely hard to prove, and probably it will remain an unsolved
> problem for a long time to come.   [Brouwer, 1975, 118]

Brouwer soon attacked this problem of the dimensional invariance
and solved it.

In his lecture Brouwer paid special attention to the multitude of
problems surrounding the attempt to classify subsets of a 2-dimen-

---

[10] Italics in original.

[11] Brouwer included merely continuous as well as *bi*continuous transformations
among the topological transformations. Cf. [Schoenflies, 1908, 149–150].

sional space, i.e., the attempt to construct a sound *analysis situs* of the plane. His critical examination of Schoenflies' attempt was fresh in his mind. He pointed out that many difficulties lay in the way of success. For example, there were only a few known topological invariants that had been defined for dimension two, such as limit point (or accumulation point), connectedness, domain, and boundary. Jordan curves and Jordan arcs were also invariants. Indeed, just about this time Brouwer had completed his elegant proof of the Jordan curve theorem, submitting it to Hilbert for *Mathematische Annalen* with his letter of October 15 (see below). In the lecture he emphasized the difficulty of proving this seemingly obvious theorem and the difficulty of proving or refuting conjectures closely related to the theorem.

He also warned about the difficulty of investigating *analysis situs* in three dimensions, saying that we could no longer be certain that a closed Jordan surface, i.e., a one-one continuous image of a spherical surface, divides 3-space into two domains [Brouwer, 1975, 119]. Yet surprisingly, not many months later he proved a quite general form of the Jordan separation theorem in n dimensions.

Brouwer's inaugural lecture and the letter to Hilbert of the same time characterize his interests in topology just before his greatest discoveries. They also reflect his entire approach to mathematics. He had arrived at the gateway to topology via other domains of mathematics, such as Lie group theory, and perceived the need to solve some of its fundamental problems. Attacking many problems derived from Schoenflies' work, he set out on a campaign of conquest. His primary aim was to solve outstanding problems and not to construct merely 'formal' theories of mathematics. Yet in the course of his successful conquest of the basic problems he was able to create a new domain of topology—virtually from nothing! Such was Brouwer's great achievement.

Brouwer's 'Proof of the Jordan curve theorem' ('Beweis des Jordanschen Kurvensatzes') [1976, 377–381] is a good example of the progress that he had already achieved in topology by the time of giving his lecture in October, 1909. In the proof he took a direct approach to the curve, rather than an approach through polygonal approximations. (Only Oswald Veblen [1905] had tried this approach before him.) It enabled him to clarify the statement of the theorem itself.

> *The one-one continuous image of a circle determines two domains in the plane and is identical with the boundary of each of these domains; in other words, it is a closed curve in the Schoenfliesian terminology.* [Brouwer, 1976, 377][12]

Brouwer used the following separation properties as the basis for his proof. Given two disjoint bounded continua (i.e., closed, thus compact, connected point sets), there is an *intermediate domain* between them (*Zwischengebiet*) that is determined by them. In the intermediate domain of two such continua a polygonal path connecting one continuum to the other determines only one domain, i.e., the path does not separate or disconnect the intermediate domain. However, in the intermediate domain of two continua two nonintersecting polygonal paths connecting one continuum to the other determine two subdomains; n nonintersecting polygonal paths connecting one continuum to the other determine n subdomains.

Having demonstrated these important separation properties, Brouwer was able to proceed to the main part of his proof. First he showed that the boundary of a domain determined by a Jordan curve is identical with the entire curve (theorem 1). This theorem accounts for his sharpening of the Jordan theorem. Then he proved that (1) a Jordan curve cannot determine more than two domains (theorem 2) and (2) a Jordan curve must determine at least two domains (theorem 3). Together these theorems completed Brouwer's elegant proof of the entire Jordan curve theorem. The main thrust of his argument was combinatorial in nature, though ostensibly he offered a proof that was set-theoretic.

Struck by its astonishing simplicity, Felix Hausdorff [1914, 354–358] incorporated Brouwer's proof in his well-known treatise, *Fundamentals of Set Theory*. Later J. W. Alexander [1920] apparently took the proof as a model for his own combinatorial proof of the Jordan curve theorem.

---

[12] Italics in original.

5. BROUWER'S DEVELOPMENT OF THE CONCEPT OF MAPPING DEGREE

We must now turn to Brouwer's work on some specific problems that caused him to fashion topological weapons powerful enough to enable him to conquer the most important problems in the topology of his day. Near the beginning of 1909 he produced two papers on related topological subjects: a paper on one-one continuous mappings of spheres into themselves and one on continuous vector fields on spheres. His investigations of these subjects spurred his development of new topological methods and results. In particular, he developed his principal new weapon for solving topological problems: the concept of degree of a continuous mapping.

The first of these papers, presented to the Royal Dutch Academy at Amsterdam on February 27, 1909, initiated a series on 'Continuous one-one transformations of surfaces in themselves' [Brouwer, 1976, 195–206]. Brouwer's investigations concerning Lie groups motivated this research. Eventually the research led to the plane translation theorem. The first paper of the series contained his earliest fixed point theorem.

> A continuous one-one transformation in itself with invariant indicatrix of a singly [i.e., simply] connected, two-sided, closed surface possesses at least one invariant point. [Brouwer, 1976, 204]

In other words, a one-one continuous orientation-preserving mapping of a surface homeomorphic to a 2-dimensional sphere to itself has at least one fixed point. Brouwer proved this important theorem by methods adapted from point set theory in the style of Schoenflies. In addition he noted that such a mapping for a 2-dimensional sphere which inverts the orientation does not necessarily have a fixed point. In contrast, the situation is exactly the reverse when we consider the 1-dimensional versions of these results [Brouwer, 1976, 205]. For a closed curve, the homeomorph of a circle (or 1-dimensional sphere), an orientation-inverting mapping always has a fixed point, whereas an orientation-preserving one will not necessarily have a fixed point. Hence, we discover that the 1- and 2-dimensional cases—very special odd and even dimensional cases—are just the opposite of one another. This distinction between odd and even dimensions was soon to become extremely important in Brouwer's program of topological research.

Just a month after submitting the paper on mappings of spheres Brouwer had a paper 'On continuous vector distributions on surfaces' [1976, 273–282] communicated to the Dutch Academy (March 27, 1909). This was the first of three papers on vector fields that he eventually published. His series of papers parallels in many ways Poincaré's great contributions on vector fields related to the qualitative theory of differential equations of the 1880s. However, Brouwer was not at first aware of Poincaré's memoirs on the subject, and indeed his viewpoint was more fully topological than Poincaré's.[13]

Brouwer's investigations into vector fields on the sphere and its homeomorphs derived from Peano's [1890a] existence theorem for differential equations. Peano's theorem can be stated in the following way. The first-order differential equation

$$\frac{dy}{dx} = f(x, y),$$

where f is a continuous function of x and y, possesses at least one integral curve or solution curve through each point (x_1, y_1). Regarding the differential equation as defining a continuous vector field in the plane or (by stereographic projection) on the sphere, we see that Peano's theorem asserts the existence of solution curves that are tangent curves to the vector distribution.

In his first paper on vector fields Brouwer used the Peano theorem and Schoenflies-style topological arguments to deduce the following result:

> A vector distribution anywhere univalent [i.e., everywhere single-valued] and continuous on a singly [i.e., simply] connected, twosided, closed surface must be zero or infinite in at least one point. [Brouwer, 1976, 279]

This result says that a continuous vector field on the homeomorphic image of a sphere must have a singularity at least one point. It is akin to a theorem of Poincaré [1881, 405] on the total number and types of singularities of a vector field on a sphere.

---

[13] Cf. [Poincaré, 1881] and [Johnson, 1981, 133, appendix A].

In his paper Brouwer noted that there is a close connection between the theorem on singularities and the fixed point theorem for mappings of spheres. He remarked:

> At first sight one might even suppose that they can be directly deduced out of each other. However this is not the case; on the contrary: they complete each other. [Brouwer, 1976, 280]

For example, if we take the singularity theorem as proved and consider a one-one continuous transformation of the sphere, we can construct a vector field that corresponds to the transformation by joining each point P to its image P' with the shortest arc of the great circle containing both points and then regarding the length and direction of the arc as the vector. However, in order that the vector field be well-defined and continuous, it is necessary to assume that no points P and P' are diametrically opposed. Yet we cannot in general assume this condition for arbitrary mappings of the sphere. Hence, the fixed point theorem does not follow from the singularity theorem. Similarly we shall fail if we attempt to deduce the singularity theorem from the fixed point theorem.

Brouwer was so perplexed by the apparent logical independence of these two theorems, so definitely related, that towards the end of 1909 he asked the advice of the eminent French mathematician Jacques Hadamard. He was undoubtedly impressed by Hadamard's mathematical abilities, especially since he had heard the Frenchman lecture at the Rome Congress of Mathematicians in April 1908. That Brouwer should now consult the distinguished Frenchman turned out to have extremely significant consequences. The resulting correspondence and contacts between the two men were decisive in the formation of Brouwer's most important topological ideas.

In a letter to Brouwer, which was probably written near the beginning of December 1909, Hadamard suggested a way of deriving the fixed point theorem from the singularity theorem (see Appendix A(1) of [Johnson, 1981, 153–154]). In his reply of December 24, 1909 Brouwer rightly drew attention to an error in Hadamard's argument (see Appendix A(2) of [Johnson, 1981, 154–155]). However, what was most significant in the exchange of letters was that Hadamard referred to Poincaré's memoirs on the curves defined by differential equations. Upon reading these, Brouwer started to investigate the concept of index for vector fields. In turn, this investigation led him to his most important topological tool: the degree of a continuous mapping.

The nineteenth-century background to Brouwer's concept of mapping degree lies in the development of the idea of circulation index or winding number and its generalizations. Cauchy, Kronecker, and Poincaré were the principal developers of this mathematical idea, though others were involved too. In the 1830s Augustin Louis Cauchy (1789–1857) explicitly defined and used the index concept. Then Leopold Kronecker (1823–1891), in papers appearing from 1869 to 1878, generalized Cauchy's concept of index in his notion of characteristic and displayed applications over a wide field. Henri Poincaré put Cauchy's index and Kronecker's characteristic to good use in his excellent memoirs on the qualitative theory of differential equations, which appeared in the 1880s. Though the topological content of the Cauchy/ Kronecker index or characteristic was not explicit in the nineteenth-century mathematical work, there emerged a well-defined concept that was ready for development in a topological direction.[14]

By the time of writing to Hadamard on December 24, 1909 Brouwer was coming to grips with Poincaré's work concerning the concept of index for vector fields. In his letter to Hadamard he rephrased the result of Poincaré applying to the ordinary 2-dimensional sphere:

> If the singular points are finite in number, then each of them possesses a finite index, and the algebraic sum of all these indices is equal to 2.[15]

Consider now a one-one continuous mapping of a sphere into itself. It is possible to devise a better vector field corresponding to the mapping. Let A be a particular point of the sphere that is not invariant and B be its image. Let M be an arbitrary point with image M'. Take a circle through M, M', and B, and define a vector at M to be the tangent to that arc MM' that does not

---

[14] Cf. [Johnson, 1981, 134–141] for more details.

[15] Cf. [Poincaré, 1881, 405].

contain B; hence, define in this way a vector distribution for the entire sphere. There are two types of singular points possible for this distribution: the points A and B and the invariant points. If there were no invariant points of the transformation, we would be contradicting the result of Poincaré, since the index of point A is -1 and that of B is $+1$; whence their sum is 0. Consequently, invariant points must exist.

Brouwer's argument in the letter to Hadamard was just a sketch. In order to spell it out more fully he needed to prove the result of Poincaré in complete topological generality. Thus his second and third communications on vector fields contained proofs of Poincaré's result [Brouwer, 1976, 283–318]. The third contained Hadamard's 'beautiful direct demonstration' ('belle démonstration directe'), as Brouwer put it in a letter of January 4, 1910 [Brouwer, 1976, 427]. It seems that Hadamard provided Brouwer with this simpler and more direct way of proving the result of Poincaré just before New Year 1910.

Freudenthal ([1975] and in [Brouwer, 1976, 420–428]) has boldly proclaimed New Year's Day 1910 as the birthday of true Brouwerian topology. This is a most appropriate date. At the time Brouwer was on a holiday in Paris, staying with his brother, Aldert, at 6 rue de l'Abbé de l'Epée in the Latin Quarter. He had arrived a day or two after Christmas 1909 and remained there until the middle of January 1910. He was excited by the new topological concepts related to index that he had been developing. Brouwer used the time in Paris to explore his new ideas, jotting many of them in a notebook labelled 'Potential Theory and Vector Analysis' ('Potentiaaltheorie en Vectoranalyse'). On the morning of New Year's Day 1910 he was able to compose a letter to Hilbert presenting a rough version of the concept of mapping degree, the central concept that he had arrived at through his feverish mental activity. Three days later on January 4 he drafted a similar letter to Hadamard [Brouwer, 1976, 421–422; 426–427]. It is not known for sure whether these letters were sent, but he was certainly ready to commit his ideas, however rough, to paper. A little later he had an opportunity to discuss his novel ideas and conjectures directly with Hadamard. Consequently, his stay in Paris was the starting point for his greatest period of topological research. The notebook and the two letters are a clear testimony to his restless activity at the time and the new directions of his research.

In the notebook, actually a student's exercise book, 'Potential Theory and Vector Analysis', the first pages on topology (7–15) are a record of Brouwer's struggle towards a general definition of index or degree in the course of his examination of related theorems about vector distributions and mappings of spheres. It is likely that he wrote these pages before he composed the letters to Hilbert and Hadamard, because the last page (15) contains a rough version of what he included in the letters, viz., a definition of degree.

First, page 7 opens with a definition of index suitable for a vector distribution on an *n*-dimensional hypersphere in terms of the *n*-dimensional solid angle which the vectors sweep out on the outer $(n-1)$-dimensional surface of a small volume. Then there follows a proof (pages 7–10) of the theorem that the total index of a continuous (tangential) vector field on an *n*-dimensional hypersphere is 0 for odd *n* and 2 for even *n*. This is a powerful generalization of the result of Poincaré. A much refined version of this proof eventually appeared in Brouwer's revolutionary paper 'On the mapping of manifolds' ('Über Abbildung von Mannigfaltigkeiten') [1976, 464–469].

After the material on vector fields in the exercise book Brouwer switched his attention to mappings, in particular, mappings of spheres and fixed point theorems (pages 10–15). The central issue is his famous general fixed point theorem, which is stated on page 12. A continuous mapping of a sphere with the interior to itself in *n*-space will always have an invariant point, but a continuous mapping of just an *n*-sphere to itself may fail to have invariant points in exactly two cases: (a) when *n* is even and the index, or degree, of the mapping is -1 and (b) when *n* is odd and the index is $+1$. On the pages surrounding this theorem (pages 11–15) all the material concerns mappings of spheres and closed surfaces. No proof of the theorem is given; instead, the concepts relating to the theorem are examined. These pages signal the very important *switch* in Brouwer's ideas from index, a concept appropriate to vector distributions, to degree, a concept more suitable to mappings. On pages 11–12 he looked at the possibility of defining the *index* of a mapping of a portion of a sphere to a sphere in terms of the algebraic number of times that a point of the target sphere is covered by an image point of the mapping, but he realized that difficulties can arise with folds and boundaries in the mapping. On subsequent pages he considered algebraic approximations to mappings of spheres and closed surfaces. Finally, on page 15 he arrived at the definition of index, now renamed *degree*, due to the polynomial functions used, that he thought was good enough to put into the letters addressed to Hilbert and Hadamard.

In a marginal note on page 10, apparently added later, Brouwer introduced the simplicial method. This became a key feature of his subsequent topological work.

An important motivation for Brouwer's new definition of degree came from the theory of functions of a complex variable and their Riemann surfaces. In the middle pages of the exercise book the main topic is mappings of spheres and algebraic approximations to them. In the crucial letters drafted for Hilbert and Hadamard Brouwer commented on mappings of 2-dimensional spheres relative to rational functions of a complex variable. He perceived that two continuous mappings of one sphere to another can be continuously transformed into one another, i.e., in modern terms, the mappings are homotopic, exactly when they have the same degree. In particular, all mappings of spheres that are of degree n can be continuously transformed into rational analytic functions of the nth degree. Hence, the nth degree rational functions (for example, $w = f(z) = z^n$), which we can represent by Riemann surfaces with n sheets, are canonical models for the nth degree mappings of spheres.[16]

Hence, at the very beginning of 1910 Brouwer was in possession of a workable definition of degree, a definition that he could then refine over the next few months. His route to the concept had been through his investigations into continuous mappings and vector fields and his concerted attempt to connect the two subjects. Hadamard had wisely referred him to Poincaré's work on index

---

[16] In modern terms we would say that the mappings of spheres of degree n all belong to the same homotopy class.

and singularities in vector fields. From this point he was impelled to transform the notion of index into his concept of degree.

Excitedly he wrote out his newly discovered version of the concept in the letters to Hilbert and Hadamard in the first days of 1910. In the letter to Hadamard he gave the definition of degree for a mapping of a sphere to itself, because he was concentrating his attention on fixed point theorems.[17] The definition reads as follows:

> To determine this *degree*, we introduce homogeneous coordinates (in the double sense) [i.e., barycentric coordinates], write x, y, z for the primitive sphere and ξ, η, ζ for the image, divide the sphere into a finite number of regions, and consider at first the transformations defined by the relations:
>
> $$\xi : \eta : \zeta = f_1(x, y, z) : f_2(x, y, z) : f_3(x, y, z),$$
>
> where f_1, f_2, f_3 are polynomials, which, moreover, can be different for the different regions of the sphere. We call this transformation a polynomial transformation. We define an indicatrix [orientation] on the sphere: then each point P of the image *in general position* will appear a number of times r_P with positive indicatrix, and a number of times s_P with negative indicatrix. One then demonstrates that $r_P - s_P$ is a constant: it is the degree of the polynomial transformation.
>
> We return to a general single-valued and continuous transformation. This can be approximated by a sequence of polynomial transformations; one demonstrates that the last ones all have the same degree: thus it is the degree of the limit-transformation.
>
> The degree is always a finite whole number, positive or negative. The degree of a *one-one* transformation is $+1$, if the indicatrix remains the same, and -1, if it is reversed. [Brouwer, 1976, 426]

Expressed in intuitive geometrical terms, the degree is the algebraic number of times that the image of the primitive sphere covers

---

[17] In the letter to Hadamard Brouwer considered only the degree of a mapping of a sphere *to itself*, while in the letter to Hilbert he considered the degree of a mapping of one sphere to another. There is a small error in my translation in [Johnson, 1981, 145] concerning this difference, which is corrected in the translation now given.

points of itself. For a mapping of one sphere to a second the idea of degree is the same; it is the algebraic count of coverings of points on the second sphere. Brouwer immediately recognized that higher-dimensional generalizations of this degree concept for 2-dimensional spheres are possible. In his letters addressed to Hilbert and Hadamard he set out the consequences of his general idea for mappings of n-spheres and their fixed points. A little later in his investigations he introduced the simplicial approach to mapping degree.

6. THE FIRST FRUITS OF BROUWER'S NEW METHODS IN TOPOLOGY

The first publications that Brouwer produced on his newly discovered concept of the degree of a mapping were 'Proof of the invariance of the dimension number' ('Beweis der Invarianz der Dimensionenzahl') and 'On the mapping of manifolds' ('Über Abbildung von Mannigfaltigkeiten') [1976, 430–445; 454–474]. Of these two revolutionary papers, the first brought in the degree concept implicitly for the proof of dimensional invariance, while the second spelled out the full details of the concept and included some important further consequences. Let us look at these first fruits of Brouwer's new ideas in topology.

As we have seen, Brouwer initially developed the concept of degree for n-dimensional generalizations of the singularity theorem for vector fields on spheres and the fixed point theorem for mappings of spheres. These results were presented as part of the second paper named above. However, armed with the degree concept at the beginning of 1910 Brouwer immediately started an assault on the refractory problem of proving dimensional invariance. As a result he submitted his 'Proof of the invariance of the dimension number' first, in June 1910, before he submitted his paper 'On the mapping of manifolds', in July 1910.

The problem of dimensional invariance fascinated him. No doubt he wanted to solve a well-known and difficult problem to show his abilities in the field of topology. Since Cantor had published in 1878 his celebrated result showing that continua of different dimen-

sion numbers can be put into one-one correspondence and Peano
had put forward in 1890 his counterintuitive example of a continuous mapping of a line segment onto a square, a 'space-filling' curve,
the problem of proving a general form of invariance of dimension
had remained open. By 1910 the dimensional invariance problem
was a very pressing one; only a few partial results had been
obtained. Brouwer vigorously attacked the problem.

The exercise book gives us evidence of Brouwer's campaign to
conquer the problem. In it he used a variety of methods to try to
prove the invariance. Probably he made these attempts during the
very first months of 1910. By March 1910 he was able to report to
Hilbert that he had arrived at a partial solution to the problem, a
solution directly related to the topological distinction between
spheres of odd and even dimensions and their transformations,
which he had discovered not long before:

> I am preparing an article for submission to the *Annalen* editorial board in
> which I solve the invariance problem of dimension insofar as I show that at
> least spaces of even and odd dimension number cannot be mapped one-one
> continuously onto one another.   [Brouwer, 1976, 429]

Actually the solution at which Brouwer hinted is complete (as
Freudenthal in [Brouwer, 1976, 424] has indicated), although
surprisingly he did not recognize this fact. Hence, he continued to
pursue the problem until he arrived at a solution that he could
accept as complete. This solution became the subject of his paper
'Proof of the invariance of the dimension number' (published in
1911) [1976, 430–445]. The paper—a mere five printed pages—
effectively swept away all previous attempts to prove dimensional
invariance. It is a masterpiece, but only the first in a series of
Brouwerian masterpieces depending on his newly discovered
methods.

We do not know how Brouwer arrived at the proof in the paper.
Although there are many attempts at proving the invariance in the
notebook and on scraps of paper, these do not provide a clear
record of progress towards the final solution. Nevertheless, among
the papers in the Brouwer Archive there is a very interesting
manuscript headed 'The invariance of the number of dimensions of

a space' ('De invariantie van het aantal dimensies eener ruimte'). This appears to be a draft of a lecture that he gave to the Dutch Mathematical Society in October 1910. The title was recorded in the *Jaarverslag* of the Society for 1910. The occasion of the lecture gave Brouwer his first opportunity to present his proof to a wider mathematical public before its publication. Undoubtedly he wanted to impress his Dutch colleagues with his new and profound result. The manuscript gives evidence of some improvements over the published paper, obviously because it was delivered four months after the submission of the paper.

Brouwer's approach to the problem of demonstrating that an *n*-dimensional manifold (or Euclidean space) and an $(m + h)$-dimensional manifold ($h > 0$) cannot be put into one-one continuous correspondence, i.e., that the manifolds are not homeomorphic, was through the proof of a lemma:

> In a *q*-dimensional manifold, if for a single-valued continuous mapping of a *q*-dimensional cube the maximum of the displacements [of the points] is less than half the side length, then there exists a concentric and homothetic cube which is contained entirely in the image set.   [Brouwer, 1976, 433]

This lemma has been called the 'Brouwer invariance principle'; it is a weak form of domain invariance.[18] The mapping indicated does not reduce the dimension of the original *q*-dimensional cube. Brouwer founded his proof of the lemma firmly on the idea of mapping degree and also on the auxiliary ideas of simplicial decomposition and the simplicial approximation of a mapping. He developed a proof full of geometrical insight.

From the invariance principle it was relatively easy for Brouwer to prove the invariance of dimension. He captured the complete invariance proof in the following theorems [Brouwer, 1976, 434].

> Theorem 1. In an *m*-dimensional manifold the one-one continuous image of an *m*-dimensional domain contains a domain in an arbitrary neighborhood of any of its points.

---

[18] Cf. [Alexandroff and Hopf, 1935, 364].

This is another weak form of domain invariance. Then we have:

> Theorem 2. An m-dimensional manifold cannot contain the one-one continuous image of a domain of higher dimension number.

> Theorem 3. In an m-dimensional manifold the one-one continuous image of a domain of lower dimension number is a nowhere dense point set.

Thus topological mappings of manifolds can neither lower nor raise the dimension numbers. Brouwer's proof of dimensional invariance stands as a marvel of elegance.

Just a month after submitting the paper on dimensional invariance Brouwer sent in his extraordinary paper 'On the mapping of manifolds' (submitted July 1910; published 1911) [Brouwer, 1976, 454–474]. This paper laid the foundations for his new methods in topology with the concept of mapping degree as the central support. The paper is closely related to Hadamard's [1910] 'Note on some applications of the index of Kronecker' ('Note sur quelques Applications de l'Indice de Kronecker'). Through their contact in Paris near the beginning of 1910 Brouwer and Hadamard influenced each other in the writing of these two excellent papers.

Hadamard's 'Note' is a magisterial presentation of the concept of Cauchy/Kronecker index or characteristic, including many consequences. It has exerted a considerable influence on later topologists.[19]

Brouwer's related paper on mappings is even more important than Hadamard's 'Note'. It included Brouwer's definition of n-dimensional manifold, his delineation of the concepts of simplicial mapping and simplicial approximation of mappings, and his fundamental definition of the degree of a continuous mapping. This cluster of basic concepts and illuminating ideas has had a profound impact on the later development of topology.

A Brouwerian *n-dimensional manifold* is a connected simplicial complex built from n-dimensional *elements,* i.e., topological images of (closed) n-dimensional Euclidean simplexes, such that the simplicial stars of the vertices have finitely many elements and form

---

[19]E.g., S. Lefschetz and J. W. Alexander.

neighborhoods of the vertices.[20] Brouwer's definition was an important step in the development of the modern concept of manifold. In order to define mapping degree, he defined *simplicial mappings,* i.e., mappings of manifolds that are piecewise or simplexwise linear. Also he defined *simplicial approximations to mappings,* i.e., linear approximations to continuous mappings. This sequence of definitions finally led to the definition of the concept of mapping degree, which was refined from the letters to Hilbert and Hadamard of January 1910. In the paper he took a global view of degree. Considering first a mapping of a closed (i.e., having finitely many elements) orientable n-dimensional manifold μ to another such manifold μ' and simplicial approximations to the mapping, he defined the *degree* for *ordinary points* P of μ' as the number p of image simplexes with a positive orientation less the number p' of image simplexes with a negative orientation covering P. This degree $p - p'$ is a well-defined integer (positive, negative, or zero), which in rough terms corresponds to the algebraic number of times that the image of μ covers points of μ'. He extended the concept of degree to mappings from closed orientable manifolds to open (i.e., having infinitely many elements) or nonorientable manifolds, but it turned out that the degree is always zero in these cases. The following theorem summarizes the material above; it brought to a close the first section of Brouwer's fundamental paper:[21]

> Theorem 1. If a twosided [i.e., orientable], closed, measured n-dimensional manifold μ is mapped to a measured n-dimensional manifold μ' in a single-valued and continuous way, then there exists a finite whole number c [the *degree* of the mapping] not changed by continuous modification of the

---

[20] Thus Brouwerian manifolds are connected n-dimensional topological manifolds having locally finite simplicial subdivisions (triangulations). A Brouwerian manifold may be compact, in which case it has a finite simplicial subdivision, or it may be just locally compact, in which case it has a locally finite simplicial subdivision. Brouwer applies the term *closed* (*geschlossen*) to a finite simplicial manifold and the term *open* (*offen*) to a merely locally finite (and infinite) simplicial manifold.

[21] The theorem refers to measured manifolds. An n-dimensional manifold is *measured* if within the elements the points are assigned barycentric coordinates, such that these form a consistent system with respect to the elements. Brouwer showed that every n-dimensional manifold can be made into a measured n-dimensional manifold.

mapping with the property that the image set of μ positively covers each subdomain of μ' c times entirely. If μ' is onesided [i.e., nonorientable] or open, then c is always zero. [Brouwer, 1976, 463]

In the second section of his paper Brouwer applied the concept of mapping degree to an analysis of continuous vector fields on n-spheres. He derived the following result on the existence of singularities.

Theorem 2. A continuous vector field on a sphere of even dimension number possesses at least one singularity point. [Brouwer, 1976, 469]

This is the porcupine theorem or hairy ball theorem for vector fields on spheres of even dimension. The theorem is not true for vector fields on spheres of odd dimension. These results flow from results on degree sum or total index concerning the vector fields.

In the third section of his paper Brouwer dealt with transformations and fixed points. He proved the following main result.

Theorem 3. A single-valued and continuous transformation of an n-dimensional sphere to itself that admits no fixed point possesses the degree -1 for even n, and the degree $+1$ for odd n. [Brouwer, 1976, 471]

From this general theorem immediately follow several consequences, the famous fixed point theorems:

Consequence 1. If the image set of a single-valued and continuous transformation of an n-dimensional sphere to itself does not lie everywhere dense in the entire sphere, then there certainly exists a fixed point.

Consequence 2. Every one-one and continuous transformation of a sphere of even dimension number to itself that can be continuously deformed into the identity certainly possesses a fixed point.

Consequence 3. Every one-one and continuous transformation of a sphere of odd dimension number to itself that can be continuously deformed into a reflection certainly possesses a fixed point. [Brouwer, 1976, 471]

Finally, considering the transformation of an n-dimensional topological simplex or element, Brouwer gave the following result—his most important fixed point theorem.

Theorem 4. A single-valued and continuous transformation of an n-dimensional element [i.e., homeomorphic image of a closed n-simplex] to itself certainly possesses a fixed point. [Brouwer, 1976, 472]

Upon the foundation of the concept of degree of a continuous mapping and closely related concepts Brouwer was able to build a sound structure of theorems. His first two papers using the new methods of topology show a tremendous range of results.

7. CONCLUSION

By the time of writing his two papers, 'Proof of the invariance of the dimension number' and 'On the mapping of manifolds', Brouwer had arrived at maturity as a topologist. He had at his disposal his most important topological methods. His subsequent work in topology is more or less a working out of the consequences of his fundamental discovery of the concept of mapping degree and the surrounding constellation of topological concepts. No doubt he had to put great effort into drawing out the consequences. Moreover, there were many surprises along the way. Still, by mid-1910 Brouwer had developed his principal topological tools for research. He could set to work on the major problems that he had already largely identified.

In the limited space of this paper it is not possible to consider the full development of Brouwer's later topological work. The stream of papers that Brouwer produced between 1910 and 1913 overflows with good ideas. I can make only a few brief remarks about the contents of his later papers in topology.

A major impulse for the way that Brouwer developed his new topological ideas further was a dispute that he had with the great French mathematician Henri Lebesgue (1875–1941) over proofs of dimensional invariance. During the summer of 1910 Otto Blumenthal (1876–1944), managing editor of *Mathematische Annalen*, met Lebesgue in Paris and made him aware of the forthcoming publication of Brouwer's proof of the invariance of dimension. Lebesgue responded with his own proof of the invariance based on his celebrated tiling or paving principle. Blumenthal placed Lebesgue's three-page sketch of his proof, written as a letter to the editor and dated October 14, 1910, just after Brouwer's proof in the *Annalen* [Lebesgue, 1911]. Thus the two proofs came out together in February 1911. Brouwer was furious when he saw

Lebesgue's proof-sketch. Though he recognized the insight provided by the tiling principle, he saw that the proof Lebesgue offered was utterly inadequate. When he called upon the Frenchman to make the proof sound, he found that Lebesgue was unwilling or perhaps even unable to do so. Instead Lebesgue responded in other ways, which annoyed Brouwer still further. In this way a bitter controversy arose between these two giants of modern mathematics. The dispute affected Brouwer's research work for several years to come. Fortunately there was a positive side to the dispute. It motivated Brouwer to produce some of his best work in topology.[22]

Brouwer produced a wealth of results during the years immediately after his breakthrough to new topological methods. In the next few paragraphs I should like to mention some of these.

He gave a proof, indeed two proofs eventually, of the invariance of an n-dimensional domain:

> In an n-dimensional manifold the one-one and continuous map of an n-dimensional domain [i.e., an open connected set] is again a domain.

A little later he took up the challenge to develop the consequences of domain invariance for the continuity method in the theory of automorphic functions.

Partly spurred by his dispute with Lebesgue, Brouwer produced a proof of the Jordan separation theorem generalized to n dimensions:

> In the n-dimensional [Euclidean] space R_n a Jordan manifold, i.e., the one-one and continuous image of a closed $(n-1)$-dimensional manifold, determines two domains, and is identical with the boundary of each of these domains.

Though the Jordan separation theorem is related to domain invariance, it is much harder to prove. Along with the theorem Brouwer developed some related results concerning Jordan manifolds.

---

[22] For a full description of the Brouwer-Lebesgue dispute and its consequences for the development of topology see [Johnson, 1981, Chapter 7] on the history of topological dimension theory.

Both Lebesgue and Brouwer tried to understand the topological concept of linking, but Brouwer gave a more rigorous working out of the concept in his notion of looping coefficient. Eventually Brouwer's and also Lebesgue's work on linking and the Jordan separation theorem took on a significant role in the later development of the celebrated Alexander duality theorem.

Brouwer also provided a sound proof of a generalization of the closed curve theorem of Schoenflies. The generalized theorem and proof had important consequences for developments in topological research during the twenties.[23]

He considered the concept of class or homotopy class with respect to his concept of degree and derived some partial results. Later Hopf derived a full characterization of homotopy classes of mappings of n-spheres by the degrees of the mappings.

As a finale to his greatest period of research in topology Brouwer [1976, 540–553] produced a second paper on the concept of dimension and dimensional invariance, 'On the natural concept of dimension' ('Über den natürlichen Dimensionsbegriff') (1913). In this paper he critically examined a definition of the concept of dimension proposed by Poincaré and offered one of his own. Then he proved dimensional invariance on the basis of his definition and a rigorous demonstration of Lebesgue's tiling principle. This is a profound work, though at first it was not as influential as it ought to have been.

There can be no doubt that Brouwer's results in topology are of immense importance. He had the vision to see the role of degree in topology, and this led him to create an entire chapter of the subject.[24]

In October 1912 Brouwer delivered his inaugural lecture as Professor in the University of Amsterdam [1975, 123–138]. In it he returned to his favored theme of intuitionism, contrasting his view with formalism. This lecture effectively marked an end to his greatest period of work in the field of topology. Over the next few

---

[23] Cf. the remarks of Freudenthal in [Brouwer, 1976, 526].

[24] For the full published details of Brouwer's most brilliant work the reader should study [Brouwer, 1976, 419–608].

years, particularly after the war, he became increasingly involved with the full development of intuitionistic mathematics. His own topological work became less important to him, though he frequently encouraged the efforts of others in the field.

Brouwer did not develop his topological work within the bounds of his intuitionistic philosophy. Nevertheless, there is a good but simple sense in which this work had an intuitionistic character. His way of doing topology stressed geometrical *construction*, geometrical thought-construction that depended upon a primitive intuition. The reader might like to examine his proofs of dimensional invariance in order to see this fact. Still, his topological work was not fully intuitionistic, because he did not show its connection with the basic intuition (referred to in Section 4) or bring it within the full strictures of intuitionistic reasoning.[25]

Above all Brouwer was a solver of problems in mathematics. In the field of topology his search for solutions to open problems was very successful. By the time of the appearance of his second paper on dimension in 1913 he had solved virtually all of the problems in topology that Schoenflies had left to him. He had executed his program of research and derived magnificent results.

Brouwer's influence in the field of topology was slow to grow. The first mathematicians to take Brouwer's ideas further were J. W. Alexander, Erhard Schmidt, Heinz Hopf, and P. S. Alexandroff. Fortunately, during the twenties and thirties Brouwer's ideas, methods, and results came to exert a profound influence. A great testament to Brouwer's topological work is the treatise by Alexandroff and Hopf, *Topology* I (*Topologie* I) [1935]. Dedicated to the great Dutchman, this work shows the marks of his topological work everywhere. In it we have evidence of an 'Age of Brouwer' in topology.

---

[25] During the twenties and later Brouwer recast some of his earlier topological proofs in intuitionistic forms.

BIBLIOGRAPHY

Alexander, James Waddell, 1920, 'A proof of Jordan's theorem about a simple closed curve,' *Annals of Mathematics* (2) 21 (1920), 180–184.
A beautiful proof of the Jordan curve theorem. Afterwards Alexander produced his celebrated duality theorem (1922).

Alexandroff, Paul (Pavel) Sergeevich and Heinz Hopf, 1935, *Topologie* I (*Topology* I), Berlin (Reprinted: New York, 1965).
A classic treatise on topology influenced both by Poincaré's work and by Brouwer's work.

Brouwer, Luitzen Egbertus Jan, 1975, *Collected Works* I, (ed.) A. Heyting, Amsterdam.

———, 1976, *Collected Works* II, (ed.) H. Freudenthal, Amsterdam.
The works of Brouwer, most of which are photographically reproduced from the original publication. The historical notes, especially those by Freudenthal, are very valuable. I have used this edition for references to Brouwer's works in this paper.

Cantor, Georg, 1878, 'Ein Beitrag zur Mannigfaltigkeitslehre' ('A contribution to manifold theory'), *Journal für die reine und angewandte Mathematik* 84 (1878), 242–258.
Cantor's proof that continua of different dimensions can be put in one-one correspondence.

———, 1932, *Gesammelte Abhandlungen mathematischen und philosophischen Inhalts* (*Collected Mathematical and Philosophical Papers*), (ed.) E. Zermelo, Berlin. (Reprinted: Hildesheim, 1962).
The collected papers of Cantor, the founder of modern set theory.

Dauben, Joseph Warren, 1979, *Georg Cantor: His Mathematics and Philosophy of the Infinite*, Cambridge, Massachusetts.
An important recent biography of Cantor. However, the older biography by Fraenkel is still very valuable; cf. [Cantor, 1932].

Freudenthal, Hans, 1975, 'The cradle of modern topology, according to Brouwer's Inedita,' *Historia Mathematica* 2 (1975), 495–502.
A valuable paper on the origins of Brouwer's most important methods in topology. Cf. Freudenthal's notes in [Brouwer, 1976].

Hadamard, Jacques, 1910, 'Note sur quelques Applications de l'Indice de Kronecker,' ('Note on some applications of the index of Kronecker'), in Jules Tannery, *Introduction à la Théorie des Fonctions d'une variable* (*Introduction to the Theory of Functions of One Variable*), Second ed., Paris, Volume II, pp. 437–477.
A classic paper on the Cauchy/Kronecker index or characteristic, closely related to Brouwer's paper 'On the mapping of manifolds' ('Über Abbildung von Mannigfaltigkeiten').

Hausdorff, Felix, 1914, *Grundzüge der Mengenlehre* (*Fundamentals of Set Theory*), Leipzig, Berlin. (Reprinted: New York, 1949, 1965).
A very influential treatise on set theory and point set topology.

Hilbert, David, 1903, 'Ueber die Grundlagen der Geometrie' ('On the foundations of geometry'), *Mathematische Annalen* 56 (1903), 381–422.
Hilbert's attempt to characterize planar geometry in a topological way using certain groups of transformations.

Johnson, Dale Martin, 1977, 'Prelude to Dimension Theory: The Geometrical Investigations of Bernard Bolzano,' *Archive for History of Exact Sciences* 17 (1977), 261–295.

———, 1979, 'The Problem of the Invariance of Dimension in the Growth of Modern Topology, Part I,' *Archive for History of Exact Sciences* 20 (1979), 97–188.

———, 1981, 'The Problem of the Invariance of Dimension in the Growth of Modern Topology, Part II,' *Archive for History of Exact Sciences* 25 (1981), 85–267. These papers constitute my history of topological dimension theory.

Jordan, Marie Ennemond Camille, 1887, *Cours d'Analyse de l'École Polytechnique.* III (*Course of Analysis at the Polytechnical School* III), Paris.

———, 1893, *Cours d'Analyse de l'École Polytechnique* I (*Course of Analysis at the Polytechnical School* I), Second ed., Paris.

Volumes from Jordan's celebrated course that contain two versions of his proof of the closed curve theorem. Incidentally, the second edition of this course is much more rigorous than the first and has been very influential on the development of modern analysis.

Lebesgue, Henri, 1911, 'Sur la non-applicabilité de deux domaines appartenant respectivement à des espaces à n et $n + p$ dimensions' ('On the nonapplicability of two domains belonging respectively to spaces of n and $n + p$ dimensions'), *Mathematische Annalen* 70 (1911), 166–168.

Paper containing Lebesgue's tiling principle.

Peano, Giuseppe, 1890, 'Sur une courbe, qui remplit une aire plane' ('On a curve that fills a planar area'), *Mathematische Annalen* 36 (1890), 157–160.

Peano's example of a 'space-filling' curve.

———, 1890a, 'Démonstration de l'intégrabilité des équations différentielles ordinaires' ('Demonstration of the integrability of ordinary differential equations'), *Mathematische Annalen* 37 (1890), 182–228.

Peano's existence theorem for differential equations.

Poincaré, Henri, 1881, 'Mémoire sur les courbes définies par une équation différentielle' ('Memoir on the curves defined by a differential equation'), *Journal de Mathématiques pures et appliquées* (3) 7 (1881), 375–422.

The first paper in a series (1881, 1882, 1885, 1886) on the qualitative theory of differential equations, a theory that Poincaré founded.

———, 1953, *Oeuvres* VI (*Works* VI), Paris.

This volume contains Poincaré's classic papers on topology, which were originally published between 1895 and 1904.

Schoenflies, Arthur, 1904, 'Beiträge zur Theorie der Punktmengen I' ('Contributions to the theory of point sets I'), *Mathematische Annalen* 58 (1904), 195–234.

———, 1904a, 'Beiträge zur Theorie der Punktmengen II' ('Contributions to the theory of point sets II'), *Mathematische Annalen* 59 (1904), 129–160.

———, 1906, 'Beiträge zur Theorie der Punktmengen III' ('Contributions to the theory of point sets III'), *Mathematische Annalen* 62 (1906), 286–328.

———, 1908, *Die Entwickelung der Lehre von den Punktmannigfaltigkeiten. Bericht, erstattet der Deutschen Mathematiker-Vereinigung. Zweiter Teil* (*The Development of the Theory of Point Manifolds. Report given to the German Mathematicians' Union. Second Part*), Leipzig.

This series of papers and this monograph contain Schoenflies' main topological work. It was severely criticized by Brouwer.

Stigt, Walter P. van, 1979, 'L. E. J. Brouwer: Intuitionism and Topology,' *Proceedings of the Bicentennial Congress, Wiskundig Genootschap*, Amsterdam, Part II, pp. 359–374.

An interesting view of the development of Brouwer's intuitionism and topological thought.

_____, 1979a, 'The Rejected Parts of Brouwer's Dissertation *On the Foundations of Mathematics*,' *Historia Mathematica* 6 (1979), 385–404.

Valuable information on the history of Brouwer's doctoral thesis.

Veblen, Oswald, 1905, 'Theory of plane curves in non-metrical analysis situs,' *Transactions of the American Mathematical Society* 6 (1905), 83–98.

Veblen's proof of the Jordan curve theorem, contained in this paper, is usually considered to be the first rigorous proof.

Willard, Stephen, 1970, *General Topology*, Reading, Massachusetts.

A standard modern text on topology.

A HOUSE DIVIDED AGAINST ITSELF: THE EMERGENCE OF FIRST-ORDER LOGIC AS THE BASIS FOR MATHEMATICS[1]

Gregory H. Moore

1. INTRODUCTION

During the nineteenth century, in a development called the arithmetization of analysis, mathematicians made precise the notions of real number and of natural number. The outcome of this research—by Cantor, Dedekind, and Weierstrass, among others—was to characterize the real numbers as a Dedekind-complete ordered field and, analogously, to characterize the natural numbers by the Peano Postulates.[2] What mattered to these mathematicians was the *uniqueness* of the structure characterized. Up to isomorphism, there existed only one Dedekind-complete ordered field and, likewise, only one structure satisfying the Peano Postulates.

In 1923 a young Norwegian mathematician, Thoralf Skolem, proposed that set theory be treated within the framework of first-order logic. One effect of this radical proposal was to ensure that the real numbers cannot be uniquely characterized by a set of axioms. During the decades that followed, mathematical logicians came to accept first-order logic as the basis for mathematics—even though Skolem [1933; 1934] had also shown that, within first-order logic, the Peano Postulates do not uniquely characterize the natural numbers.

Skolem did not consciously set out to undo the work of Cantor, Dedekind, Peano, and Weierstrass. Nevertheless, to a mathematician who is not a logician, it may seem odd that mathematical logicians chose as the basis for mathematics a system of logic that cannot characterize any infinite system. Yet this is precisely what occurred with the adoption of first-order logic.

The present chapter describes how first-order logic emerged and how, despite the opposition of eminent mathematicians like Zermelo, it came to be accepted as the proper basis for mathematics by the vast majority of mathematical logicians. It will also be seen that first-order logic emerged from two stronger kinds of logic.

2. WHAT IS FIRST-ORDER LOGIC?

A formal language has certain *logical symbols*: propositional connectives ("or," "not," "and," "if...then"), quantifiers ("there exists," "for all"), and variables for individuals. Such a language also has certain *nonlogical symbols* relative to the particular theory being formalized: constant symbols, function symbols, and relation symbols. These nonlogical symbols are then interpreted, relative to a given universe A of objects, by specifying for each constant symbol a particular object in the universe, for each n-ary function symbol a particular function from A^n to A, and for each n-ary relation symbol a particular relation on A^n. In brief, this is the basic structure of first-order logic.

Besides the individual variables of first-order logic, the formal language of *second-order logic* has additional variables, namely, function variables and relation variables. Thus second-order logic permits quantifiers that range over sets (or functions or relations) rather than merely over individuals (the members of the universe in the given interpretation). These quantifiers are then interpreted as varying over all sets (or all functions or all relations) in the given universe. It is the existence of such second-order quantifiers that makes it possible for the usual axiomatizations of the real numbers and the natural numbers to be *categorical* (that is, any two models of the axioms are isomorphic).

Consider number theory: In first-order logic, we have one constant symbol and one unary function symbol (to be interpreted as zero and as the successor function, respectively). The Peano postulate for mathematical induction is a second-order axiom, since it uses a quantifier ranging over all subsets of the natural numbers. More precisely, it says that, for any set P, if 0 belongs to P and, for every n, if n belongs to P then $n + 1$ belongs to P, then every natural number belongs to P. In first-order logic this axiom cannot be expressed; instead, it becomes an axiom schema, obtained by replacing the subset quantifier by the countably many particular subsets definable in the system.[3] Since the second-order axiom has a quantifier ranging over *uncountably* many subsets, this axiom is clearly stronger than the corresponding first-order axiom schema.

Certain concepts and distinctions (fundamental both to first-order logic and to the modern mathematician's informal logic) were not always clearly formulated or consistently observed during the nineteenth and early twentieth centuries. Chief among these is the distinction between *syntax* (including such notions as formal language, formula, proof, and consistency) and *semantics* (including such notions as truth, model, and satisfiability). Finally, many logicians of that time, but by no means all of them, assumed that logic has only formulas and proofs of finite length—thus ruling out what is now called *infinitary logic*, which began to develop vigorously only in the 1950s.

3. BOOLE: THE EMERGENCE OF MATHEMATICAL LOGIC

Before the nineteenth century, few mathematicians investigated formal logic. Since the time of Aristotle, who established the canons of formal logic in the fourth century B.C., logic was regarded as a part of philosophy. Those mathematicians who thought about logic, such as Leibniz and Euler, had strong philosophical interests. Nevertheless, mathematicians engaged in a kind of reasoning that presumably could be treated as a part of formal logic.

Until the twentieth century, there was no reason to think that the kind of logic that one used would affect the mathematics that one did. Indeed, around 1850 the Aristotelian syllogism remained the archetype of *all* reasoning. When mathematical logic began in 1847, with the publication of George Boole's *The Mathematical Analysis of Logic*, Aristotelian logic was treated as one interpretation of a particular algebraic system. In that context, logic was a part of mathematics. Logic itself, on the other hand, continued to be understood by Boole as "the laws of thought." His research, which led to the formulation of Boolean algebra, was on the border of philosophy, psychology, and mathematics:

> The design of the following treatise is to investigate the fundamental laws of those operations of the mind by which reasoning is performed; to give expression to them in the symbolical language of a Calculus, and upon this foundation to establish the science of Logic and construct its method; ... and, finally, to collect from the various elements of truth brought to view in the course of these inquiries some probable intimations concerning the nature and constitution of the human mind. [Boole, 1854, 1]

Boole set up an *uninterpreted calculus* of symbols and operations, and then gave this calculus various *interpretations*—including one in terms of classes and one in terms of propositions. This calculus was based upon conventional algebra suitably modified for use in logic. He pursued the analogy with algebra by introducing formal symbols for the four arithmetic operations of addition, subtraction, multiplication, and division as well as by using functional expansions. Deductions took the form of equations and of the transformation of equations. His symbol v played the role of an indefinite class, serving in place of an existential quantifier [Boole, 1854, 124].

In time, a modified version of Boole's logic would become the lowest level of logic and would be called *propositional logic*: the level of logic having propositional connectives but no quantifiers. Yet when Boole wrote, his system functioned as all of logic, since within this system Aristotelian syllogistic logic could be interpreted.

4. PEIRCE AND FREGE: THE NOTION OF QUANTIFIER

A kind of logic adequate for mathematics emerged late in the nineteenth century when two related developments took place: first, relations and functions were introduced into symbolic logic; second, the notion of quantifier was disentangled from that of propositional connective and was given an appropriate symbolic representation. These developments were brought about by two mathematicians having a strong philosophical bent—Charles Sanders Peirce in the United States and Gottlob Frege in Germany.

Yet the notion of quantifier has an ancient origin. Aristotle made the notions of "some" and "all" central to logic by formulating the syllogism—a very restricted mode of deduction. An example of one of the more complicated syllogisms considered by Aristotle is the following argument: Suppose that every A has property B and that some C does not have property B; then it follows that some C is not an A. Boole's logic was capable of handling such simple uses of quantifiers by, in effect, interpreting A, B, and C as classes. What happened during the 1870s and 1880s, thanks to Peirce and Frege, was that the notion of quantifier was *distinguished* from the Boolean connectives on the one hand and from the notion of predicate (or relation) on the other.

Peirce's contributions to logic fell squarely within the Boolean tradition. In [1865] Peirce modified Boole's system, reinterpreting Boole's + (logical addition) as union in the case of classes and as inclusive "or" in the case of propositions. (Boole had defined $A + B$ only when A and B are disjoint.)

Five years later Peirce treated the notion of relation, which Augustus De Morgan had introduced into formal logic [De Morgan, 1859], and enlarged Boole's system to include this notion:

> Boole's logical algebra has such singular beauty, so far as it goes, that it is interesting to inquire whether it cannot be extended over the whole realm of formal logic, instead of being restricted to that simplest and least useful part of the subject, the logic of absolute terms, which, when he wrote [1854], was the only formal logic known. [Peirce, 1870, 317]

In particular, Peirce studied the laws of the relative product (which generalizes the notion of composition of functions) and the converse of a relation. When Peirce left for Europe in June of 1870, he took along offprints of this article and delivered one to De Morgan [Fisch, 1984, xxxiii]. Unfortunately, De Morgan was already in the decline that led to his death the following March. Peirce did not find a better reception when he gave an offprint to Stanley Jevons, who had elaborated Boole's system in England. For in a letter of August 1870 to Jevons, whom Peirce described as "the only active worker now, I suppose, upon mathematical logic," it is evident that Jevons rejected Peirce's extension of Boole's system to relations [Peirce, 1984, 445]. Nevertheless, Peirce's ideas on relations were later widely accepted by mathematical logicians.

By applying the operations of union and intersection to relations, Peirce [1883] developed the notion of quantifier as something distinct from the Boolean connectives. As an example, he let l_{ij} denote the relation stating that i is a lover of j. "Any proposition whatever is equivalent," he explained,

to saying that some complexus of aggregates and products of such numerical coefficients is greater than zero. Thus,

$$\sum_i \sum_j l_{ij} > 0$$

means that something is a lover of something; and

$$\prod_i \sum_j l_{ij} > 0$$

means that everything is a lover of something. We shall, however, naturally omit, in writing the inequalities, the > 0 which terminates them all; and the above two propositions will appear as

$$\sum_i \sum_j l_{ij} \quad \text{and} \quad \prod_i \sum_j l_{ij}.$$

[Peirce, 1883, 200–201]

When he next discussed quantifiers, he treated them in two ways that strongly influenced later research. First, he defined the quantifiers (within what he called the "first-intentional logic of relatives") in a manner that emphasized their analogy with arithmetic:

> Here, in order to render the notation as iconical as possible, we may use Σ for *some*, suggesting a sum, and Π for *all*, suggesting a product. Thus $\Sigma_i x_i$ means that x is true of some one of the individuals denoted by i or
>
> $$\sum_i x_i = x_i + x_j + x_k + \text{etc.}$$
>
> In the same way,
>
> $$\prod_i x_i = x_i x_j x_k \text{ etc.}$$
>
> If x is a simple relation, $\Pi_i \Pi_j x_{ij}$ means that every i is in this relation to every j, $\Sigma_i \Pi_j x_{ij}$ that some one i is in this relation to every j.... It is to be remarked that $\Sigma_i x_i$ and $\Pi_i x_i$ are only *similar* to a sum and a product; they are not strictly of that nature, because the individuals of the universe may be innumerable. [Peirce, 1885, 194–195]

Thus Peirce regarded a formula with an existential quantifier, "for some i, $A(i)$," as an infinitely long formula,

$$A(i) \text{ or } A(j) \text{ or } A(k) \text{ or} \ldots,$$

where i, j, k, etc. were names for all the individuals in the given universe of discourse. An analogous connection held between "for all i, $A(i)$" and

$$A(i) \text{ and } A(j) \text{ and } A(k) \text{ and} \ldots.$$

Here, and in the writings of those who later followed Peirce's approach (such as Ernst Schröder and Leopold Löwenheim), the syntax was not distinguished from the semantics, for there was given in advance a particular domain (a semantic concept) over which the quantifiers were to range. When the domain was infinite, it was natural to think of quantifiers in such an infinitary fashion, since, for a finite domain of elements i_1 to i_n, the formula "for all i, $A(i)$" reduced to

$$A(i_1) \text{ and } A(i_2) \text{ and} \ldots \text{and } A(i_n),$$

and "for some i, $A(i)$" reduced to

$$A(i_1) \text{ or } A(i_2) \text{ or} \ldots \text{or } A(i_n).$$

Hence, unlike the logic of Peano for example, Peirce's logic was not restricted to formulas of finite length.

A second way in which Peirce's treatment of quantifiers significantly affected later developments occurred in what he called "second-intensional logic." This kind of logic, which permitted predicates to be quantified, was a version of second-order logic. Peirce used this logic to define the notion of identity (a definition possible in second-order logic but not, in general, in first-order logic):

> Let us now consider the logic of terms taken in collective senses [second-intensional logic]. Our notation...does not show us even how to express that two indices, i and j, denote one and the same thing. We may adopt a special token of second intention, say 1, to express identity, and may write $1_{ij} \ldots$. And identity is defined thus:
>
> $$1_{ij} = \prod_k \left(q_{ki} q_{kj} + \bar{q}_{ki} \bar{q}_{kj} \right).$$
>
> That is, to say that things are identical is to say that every predicate is true of both or false of both. [Peirce, 1885, 199]

In effect, Peirce used a version of Leibniz' principle of the identity of indiscernibles in order to give a second-order definition of identity.

What Peirce glimpsed of second-order logic was minimal. He rarely returned to his second-intensional logic, though he used it, in a 1900 letter to Cantor, to quantify over relations while defining the less-than relation for cardinal numbers [Peirce, 1976, 776]. Moreover, he never applied his logic in detail to mathematical problems, except for the beginnings of cardinal arithmetic [Peirce, 1885]—an omission that contrasts sharply with Frege.

Frege's logic differed substantially both from Boole's system and from first-order logic. The *Begriffsschrift* [1879], Frege's first publication on logic, was influenced by two of Leibniz' ideas: a *calculus ratiocinator* (a formal calculus of reasoning) and a *lingua characteristica*, (a universal language). As a step toward such a *lingua characteristica* Frege introduced a formal symbolic language on which to found arithmetic. Unlike Boole, Frege allowed the interaction between logic and mathematics to proceed in both directions. From mathematics Frege borrowed the notions of function and argument to replace the traditional logical notions of predicate and subject, and then he used the resulting logic as a basis for constructing arithmetic.

Independently of Peirce, Frege introduced his universal quantifier in such a way that functions could be quantified as well as arguments, and he used such function quantifiers to state the principle of mathematical induction [Frege, 1879, Sections 11 and 26]. In order to develop the general properties of sequences, Frege both wanted and believed that he needed a logic at least as strong as what was later called second-order logic.

Frege's notions of logical function and argument evolved further in his book, *Foundations of Arithmetic*, where he wrote of making "one concept fall under a higher concept, so to say, a concept of second order" [Frege, 1884, Section 53]. This second-order concept was analogous to a property of properties of individuals.[4]

Frege's most detailed treatment of such functions is found in his two-volume work *Fundamental Laws of Arithmetic* [1893; 1903]. There quantification over his functions of second order (in effect, second-order logic) played a central role. Of the six axioms on which his logic was based, two unequivocally belonged to second-order logic:

(1) If $a = b$, then for every property f, a has the property f if and only if b has f.

(2) If a property $F(f)$ of properties f holds for every property f, then $F(f)$ holds for any particular property f.

Frege's system would have taken a fundamentally different form if he had wanted to dispense with second-order logic. At no point did he give any indication of wishing to do so. Furthermore, given his logical conceptions, he could not have defined the notion of cardinal number as he did, deriving it from logic, without the use of second-order logic.

In the *Fundamental Laws* Frege introduced a hierarchy of levels of quantification. Not only did he consider propositional functions of an individual variable x (first-order propositional functions) and propositional functions of propositional functions of x (second-order propositional functions) but he also considered third-order propositional functions.

Yet he introduced his axioms as second-order, not third-order, propositional functions. The reason that, in effect, Frege developed a second-order logic rather than a third-order or still higher-order logic was that, in his system, second-level concepts could be represented by their extensions (*i.e.*, as sets) and thereby appear in propositions as objects [Frege, 1893, 42]. However, this treatment, when combined with his fifth axiom (a form of the Principle of Comprehension), made his system contradictory—as Russell was to inform him (using Russell's Paradox) in 1902.

Frege did not separate the first-order part of his logic from the rest of that logic. Nor could he have done so without doing violence to his principles. Such a separation only occurred decades later, and then within the Boole-Peirce-Schröder tradition.

5.   SCHRÖDER: QUANTIFIERS IN THE ALGEBRA OF LOGIC

Ernst Schröder, who began his work in logic within the Boolean tradition [Schröder, 1877], was not then acquainted with Peirce's contributions to logic. On the other hand, Schröder soon learned of Frege's *Begriffsschrift* and gave it a lengthy review [Schröder, 1880]. This review praised the *Begriffsschrift* and added that it promised to help advance Leibniz' goal of a universal language. However, Schröder then criticized Frege for failing to take account of Boole's contributions, arguing that what Frege had done could be done more clearly by means of Boole's notation.

Three years later Frege replied to Schröder, emphasizing the differences between Boole's symbolic language and his own: "I did not wish to represent an abstract logic by formulas but to express meaning by written signs in a more exact and clear fashion than is possible by words" [Frege, 1883, 1]. Frege's remark indicates that in his logic, but not in Boole's, the propositions carried an intended interpretation. Frege also stressed that his notation allowed a universal quantifier to apply only to a *part* of a formula, whereas Boole's notation did not permit such a restriction. It was this point which Schröder had overlooked in his review and which, following Peirce, he would incorporate into his later work: the separation of quantifiers from the Boolean connectives.

Schröder adopted this separation in the second volume of his *Lectures on the Algebra of Logic* [1891], a three-volume study of logic (within the tradition of Boole and Peirce) that was rich in algebraic techniques applied to semantics. In the first volume [1890] Schröder discussed the "identity calculus," which was essentially Boolean algebra, and three related subjects: the propositional calculus, the calculus of classes, and the calculus of domains. When, in 1891, he treated Peirce's notation for quantifiers, he used it in a second-order fashion to quantify over all subdomains of a given domain (or manifold) called 1: "In order to express that a proposition concerning a domain x holds... *for every domain x* (in our manifold 1), we shall place the sign Π_x before [the proposition]..." [Schröder, 1891, 26]. He insisted that there is no manifold 1 containing all objects, since otherwise a contradiction would result [Schröder, 1890, 246]—a premonition of the later set-theoretic paradoxes.

Schröder's third volume [1895], devoted to the "algebra and logic of relations," contained several kinds of infinitary and second-order propositions. He used second-order logic as a tool in "elimination problems," where the goal was to solve a logical equation for a given variable. He stated a second-order proposition that he could have taken to be the definition of identity (following Peirce), but did not. However, in what he called "a procedure that possesses a certain boldness," he considered an infinitary proposition that had a universal quantifier for uncountably many (in fact, continuum many) variables [Schröder, 1895, 511, 512]. Finally, in order to move an existential quantifier to the left of a universal quantifier, he introduced a universal quantifier subscripted with relation variables (and so ranging over them), and then expanded this second-order quantifier into an infinite product of first-order quantifiers, one for each individual in the given infinite domain.[5] This general procedure would play a fundamental role in the proof that Löwenheim was to give in 1915 of Löwenheim's Theorem (see Section 9).

6. HILBERT: EARLY RESEARCHES ON FOUNDATIONS

During 1898-1899 David Hilbert gave a course of lectures at Göttingen on the foundations of Euclidean geometry, soon publishing them as a book. At the beginning of this book, which became the textbook of the modern axiomatic method, he stated his purpose:

> The following investigation is a new attempt to establish for geometry a system of axioms that is *complete* and *as simple as possible*, and to deduce from these axioms the most important theorems of geometry in such a way that the significance of the different groups of axioms and the scope of the consequences to be drawn from the individual axioms is brought out as clearly as possible. [Hilbert, 1899, 1]

Hilbert did not specify what "complete" meant in this context until a year later, when he remarked that the axioms for geometry are complete if all the theorems of Euclidean geometry are deducible from the axioms [Hilbert, 1900, 181]. Apparently what he meant was that all *known* theorems be so deducible.

On December 27, 1899 Frege began corresponding with Hilbert about the foundations of geometry. Frege had read Hilbert's book but found its approach odd. Indeed, Frege insisted on the traditional view of geometric axioms whereby they were justified by geometric intuition. Replying on December 29, Hilbert proposed a more modern view whereby an axiom system only determines up to isomorphism the objects described by it:

> You write: "I call axioms propositions that are true but are not proved because our knowledge of them flows from a source very different from the logical source, a source which might be called spatial intuition. From the truth of the axioms it follows that they do not contradict each other." I found it very interesting to read this sentence in your letter, for as long as I have been thinking, writing, and lecturing on these things, I have been saying the exact opposite: If the arbitrarily given axioms do not contradict each

other with all their consequences, then they are true and the things defined by the axioms exist. For me this is the criterion of truth and existence. (Hilbert in [Frege, 1980, 39–40])

Hilbert returned to this theme repeatedly in later years.

Frege, responding on January 6, 1900, objected vigorously to Hilbert's assertion that the consistency of an axiom system implies the existence of a model of the system. The only way to prove the consistency of an axiom system, Frege insisted, is to give a model. Furthermore, he believed that the crux of Hilbert's "error" was in confusing first-order and second-order concepts.[6] For Frege, existence was a concept of second order, and it is precisely here that his system of logic as found in the *Fundamental Laws* differs from modern second-order logic.

An important axiom was not present in the first edition of Hilbert's book but was added later. In the first edition his Axiom Group V consisted merely of the Archimedean Axiom. He then used a quadratic field to establish the consistency of his system, stressing that this proof required only a denumerable set.

When the French translation of his book came out in 1902, he included a new axiom that differed fundamentally from his other axioms and that would stimulate him to try to establish the consistency of a non-denumerable set, the real numbers:

> Let us note that to the five preceding groups of axioms we may still adjoin the following axiom which is not of a purely geometric nature and which, from a theoretical point of view, merits particular attention.

Axiom of Completeness

> To the system of points, lines, and planes it is impossible to adjoin other objects in such a way that the system thus generalized forms a new geometry satisfying all of the axioms in groups I–V.   [Hilbert, 1902, 25]

The motivation behind the Axiom of Completeness, an axiom which is false in first-order logic but possible in second-order logic, was the desire to ensure that every interval on a line contains a limit point. Yet Hilbert had some initial reservations, since he added: "In the course of the present work we have not used this 'Axiom of Completeness' anywhere" [Hilbert, 1902, 26]. When the second German edition of his book appeared a year later [1903], his reservations had lessened, and he designated the Axiom of Completeness as Axiom V2. So it remained in the seven editions published during his lifetime. This axiom gave rise to an extensive literature.

The original version of the Axiom of Completeness, which referred to the real numbers rather than to geometry, stated that the real numbers cannot be extended to a larger Archimedean ordered field. This version formed part of Hilbert's [1900] axiomatization of the real number system. In particular, he asserted that his Axiom of Completeness implies the Bolzano-Weierstrass Theorem and that his system characterizes the usual real numbers. It is not clear why he formulated this axiom as an assertion about maximal models of an axiom system rather than in more mathematically conventional terms (*e.g.*, the existence of a least upper bound for every bounded set).

That same year Hilbert gave his famous lecture on "Mathematical Problems" at the International Congress of Mathematicians held in Paris. As his second problem, he proposed that one prove the consistency of his axioms for the real numbers. At the same time he emphasized the usefulness of the axiomatic method, his belief that every well-formulated mathematical problem can be solved, and his conviction that the consistency of a set S of axioms implies the existence of a model for S [Hilbert, 1900a, 264–266]. When he gave this address, his view that consistency implies existence was only an article of faith—albeit one to which Poincaré subscribed as well [Poincaré, 1905, 819]. Yet in 1930 Gödel was to turn this article of faith into a theorem, in fact into the version of his Completeness Theorem (for first-order logic) stating that an axiom system S has a model if and only if S is consistent.

In 1904, when addressing the International Congress of Mathematicians at Heidelberg, Hilbert was still concerned with the foundations of the real number system. But his attention turned, first of all, to providing a foundation for the positive integers. While analyzing Frege's work, he discussed the paradoxes of logic and set theory for the first time in print. To Hilbert these paradoxes and contradictions indicated that "the conceptions and research methods of logic, conceived in the traditional sense, do not measure up to the rigorous demands that set theory makes"

[Hilbert, 1905, 175]. Consequently, he wished to circumvent such paradoxes in the foundations of arithmetic. His method for doing so, however, separated him sharply from Frege:

> Yet if we observe attentively, we realize that in the traditional treatment of the laws of logic certain fundamental notions from arithmetic are already used, such as the notion of set and, to some extent, that of number as well. Thus we find ourselves on the horns of a dilemma, and so, in order to avoid paradoxes, we must simultaneously develop both the laws of logic and of arithmetic to some extent. [Hilbert, 1905, 176]

Hilbert excused himself from giving more than an indication of how to carry out such a simultaneous development, but what he did say indicates how he viewed mathematical logic in 1904. For the first time his investigations relied on a formal language. Within that language his quantifiers were in the Peirce-Schröder tradition, although he did not cite those authors. Indeed, he regarded "for some x, $A(x)$" merely as an abbreviation for the infinitary formula

$$A(1) o . A(2) o . A(3) o . \ldots,$$

where o. stood for "oder" (or), and analogously for the universal quantifier with respect to "und" (and) [Hilbert, 1905, 178]. Likewise, he followed Peirce and Schröder, as well as the geometric tradition, by letting his quantifiers range over a *fixed* domain, in this case the positive integers. As axioms he took the Peano Postulates—except that the principle of mathematical induction was omitted.

Hilbert's chief concern here was to show the consistency of his axioms for the positive integers. He did so by finding a combinatorial property that held for all theorems but did not hold for a contradiction. This marks the beginning of what, over a decade later, would become his proof theory.

Thus, in 1904, Hilbert's logic embodied some elements of first-order logic but not others. Above all, his use of infinitary formulas and his restriction of quantifiers to a fixed domain differed fundamentally from first-order logic. The principal advance came when he began to apply metamathematical notions to logic itself. In this he proceeded purely *syntactically*, and another decade would pass before similar considerations were applied, by Leopold Löwenheim, to the *semantical* side of logic.

Meanwhile, the concept of the categoricity of an axiom system emerged from the work of Edward Huntington and Oswald Veblen, both of whom belonged to the group of mathematicians called the American Postulate Theorists. Huntington, while giving an essentially second-order axiomatization of the real numbers by means of sequences, introduced the term "sufficient" to mean that "there is essentially *only one* such assemblage [set] possible" that satisfies a given set of axioms [Huntington, 1902, 264], and he made it clear later in his article that his term meant that any two models are isomorphic.

In 1904 Veblen, while investigating the foundations of geometry in the tradition of Peano, discussed this terminology. John Dewey had suggested to Veblen the use of the term "categorical" for an axiom system such that any two of its models are isomorphic. Veblen cited Hilbert's axiomatization of geometry (including the Axiom of Completeness) as such a categorical system, and added that, for a categorical system, "the validity of any possible statement in these terms is therefore completely determined by the axioms; and so any further axiom would have to be considered redundant, even were it not deducible from the axioms by a finite number of syllogisms" [Veblen, 1904, 346].

Edward Huntington quickly adopted the term categorical and made a further observation:

> In the case of any categorical set of postulates one is tempted to assert the theorem that if any proposition can be stated in terms of the fundamental concepts, either it is itself deducible from the postulates, or else its contradictory is so deducible; it must be admitted, however, that our mastery of the processes of logical deduction is not yet, and possibly never can be, sufficiently complete to justify this assertion.   [Huntington, 1905, 210]

Huntington was convinced that every categorical axiom system is deductively complete, *i.e.*, every sentence expressible in the system is provable or disprovable. On the other hand, Veblen was aware of the possibility that in a categorical axiom system there might exist propositions *true* in the only model of the system but *unprovable* in the system. In 1931 Gödel's Incompleteness Theorem would establish that this possibility was realized, in second-order logic, for *every* categorical axiom system rich enough to include the arithmetic of the natural numbers.

7. PEANO AND RUSSELL

In [1888] Giuseppe Peano began his logical investigations by describing the work of Boole [1854] and Schröder [1877]. Indeed, in a letter to Peano (probably written in 1894), Frege described Peano as a follower of Boole, but one who had gone further by introducing a symbol for generalization.[7] This symbol first appeared in 1889, where it took the form $a \supset_{x,y,\dots} b$ and meant "whatever x, $y,\dots$ may be, b is deduced from a." Thus Peano, independently of Frege and Peirce, put forward the notion of universal quantifier, but did not separate it completely from the symbol for implication [Peano, 1889, Section II]. In Peano's work, however, there is no indication of levels of logic. In particular, he expressed "x is a positive integer" by $x \epsilon N$ and so felt no need to quantify over predicates such as N. Yet his axiomatization for N, later called the Peano Postulates, was essentially a second-order axiomatization, since it included the principle of mathematical induction.

In 1900 Bertrand Russell adopted and extended Peano's formal language, which thereby achieved an influence and a longevity denied to Frege's.

When he became an advocate of Peano's logic, Russell entered a new phase of his development. In contrast to his earlier views, Russell accepted Cantor's transfinite ordinal and cardinal numbers. Then, in May 1901, he discovered Russell's Paradox and, after trying sporadically to solve it for a year, wrote about it to Frege on June 16, 1902. Frege was devastated by this discovery. Although his original [1879] system of logic was not threatened, he saw that the system developed in the *Fundamental Laws* [1893] was in grave danger. For there he permitted a set ("a range of values," in his words) to be treated as an object and, consequently, to be the argument of a first-order function; in this way Russell's Paradox arose in his system.

Less than two months later, on August 8, Russell wrote Frege a letter containing the first version of the theory of types—Russell's solution to the paradoxes of logic and set theory. Russell used the language of Frege's theory:

> The contradiction [Russell's Paradox] could be resolved with the help of the assumption that ranges of values are not objects of the ordinary kind; *i.e.*, that $\phi(x)$ needs to be completed (except in special circumstances) either by an object or by a range of values of objects [set of objects] or a range of values of ranges of values [set of set of objects], etc. This theory is analogous to your theory about functions of the first, second, etc., order. (Russell in [Frege, 1980, 144])

This passage indicates that the seed of the theory of types grew directly from the soil of Frege's logic as found in the *Fundamental Laws*. Philip Jourdain later asked Frege (in a letter of January 15, 1914) whether Frege's theory was the same as Russell's theory of types. In a draft of his reply to Jourdain (on January 28), Frege answered a qualified yes:

> Unfortunately I do not understand the English language well enough to be able to say definitely that Russell's theory (*Principia Mathematica* I, 54ff) agrees with my theory of functions of the first, second, etc., order. It does seem so. [Frege, 1980, 78]

In 1903 Russell's Paradox appeared in print, in the second volume of Frege's *Fundamental Laws* [1903, 253] and in Russell's *Principles of Mathematics*. Frege dealt only with Russell's Paradox while Russell also discussed in detail the paradox of the largest cardinal and the paradox of the largest ordinal. In an appendix to his book, Russell offered a preliminary version of the theory of types as a way to resolve these paradoxes, but he remained uncertain whether this theory would actually eliminate *all* paradoxes [Russell, 1903, 528].

After several detours through other ways of eliminating paradoxes from logic, Russell [1908] wrote an exposition of his mature theory of types, the basis for *Principia Mathematica*. A type was defined to be the *range of significance* of some propositional function. The first type consisted of the individuals. The second type was that of what he called "first-order propositions": those propositions whose quantifiers range only over the first type, that is, over individuals. The third logical type consisted of second-order propositions: those propositions whose quantifiers range only over the first or second types (*i.e.*, over individuals or first-order propositions). Analogously he defined a type for each finite index. At the same time he introduced a parallel hierarchy of propositional functions, and avoided introducing classes by using such propositional functions instead. Like Peirce, he defined the identity of individuals x and y by the clause that every first-order proposition holding for x also holds for y [Russell, 1908, Sections IV–VI].

Thus, outlined in [Russell, 1908] and developed in detail in *Principia Mathematica* [Whitehead and Russell, 1910], the theory of types included first-order logic, second-order logic, and so on. However, the first-order logic that it contained differed from modern first-order logic in that a proposition about classes of classes could not be treated in first-order logic. This privileged position of the membership relation was later attacked by Skolem (see Section 10).

For Russell and Whitehead (as for Frege but not Boole), logic was a foundation for all of mathematics. Consequently, Russell and Whitehead considered it impossible to stand outside of logic and thereby to study it as a system (as one might, *e.g.*, study the real numbers). Furthermore, they lacked any conception of a *metalanguage*, and would surely have rejected such a conception if it had been proposed to them. In this vein, they explicitly denied the possibility of independence proofs for their axioms, and they added that mathematical induction could not be used to prove theorems *about* their system of logic [Whitehead and Russell, 1910, 95, 135].

Metatheoretic research about the theory of types had to come from those schooled in a different tradition. When the consistency of the simple theory of types (without the Axiom of Infinity) was eventually proved, it was done by Gerhard Gentzen [1936], a member of Hilbert's school, and not by someone within the logicist tradition of Frege, Russell, and Whitehead.

Russell and Whitehead, like Frege, lacked the notion of *model* or interpretation. Instead, they employed the genetic method of constructing, for example, the positive integers, rather than using the axiomatic approach of Peano or Hilbert. Finally, Russell and Whitehead shared with many other logicians the tendency to blur syntax and semantics, as when they stated their first axiom that "anything implied by a true elementary proposition is true" [Whitehead and Russell, 1910, 98].

During the 1920s, and to a lesser extent thereafter, *Principia Mathematica* became the basis for research in mathematical logic.

8. HILBERT: LATER FOUNDATIONAL RESEARCH

Hilbert did not abandon foundational questions after his 1904 Heidelberg lecture, though he published nothing on them for a decade. He gave lecture courses at Göttingen on such questions repeatedly—in 1905, 1908, 1910, and 1913. Yet it was in the course given in 1917–1918, "Principles of Mathematics and Logic," that he first exhibited his mature conception of logic.

The course began shortly after Hilbert delivered a lecture at Zurich in September 1917, which stressed the role of the axiomatic method in mathematics and physics. In that lecture, returning to a theme from his Paris address [Hilbert, 1900a], he noted how the consistency of several axiomatic systems (such as that for geometry) had been reduced to a more specialized axiom system (such as that for the real numbers, reduced in turn to the axioms for the positive numbers and those for set theory). Hilbert concluded by stating that, "this full-scale undertaking of Russell's to axiomatize logic can be seen as the crowning achievement of axiomatization" [Hilbert, 1918, 412].

The 1917 course treated the axiomatic method in two disciplines: geometry and mathematical logic. Influenced by *Principia Mathematica*, Hilbert regarded propositional logic as a separate level of logic. But he deviated from Russell and Whitehead by offering a proof for the consistency of propositional logic. Turning to first-order logic, which he called the "functional calculus," he stated primitive symbols and axioms for it. After developing it at length, he stated:

> With what we have considered thus far [first-order logic], foundational discussions about the calculus of logic come to an end—if we have no other goal than formalizing logical deduction. We, however, are not content with this application of symbolic logic. We wish not only to be in a position to develop individual theories from their principles purely formally but also to

make the foundations of mathematical theories themselves an object of investigation—to examine in what relation they stand to logic and to what extent they can be obtained from purely logical operations and concepts. To this end the calculus of logic must serve as our tool.

Now if we make use of the calculus of logic in this sense, then we will be compelled to extend in a certain direction the rules governing the formal operations. In particular, while we previously separated propositions and [propositional] functions completely from objects and, accordingly, distinguished the signs for indefinite propositions and functions rigorously from the variables, which take arguments, now we permit propositions and functions to be taken as logical variables in a way similar to that for proper objects, and we permit signs for indefinite propositions and functions to appear as arguments in symbolic expressions.   [Hilbert, 1917, 188]

Here Hilbert argued for a logic at least as strong as second-order logic. But his views did not appear in print until his book *Principles of Mathematical Logic*, written jointly with Wilhelm Ackermann, was published in 1928. In that book, which consisted largely of a revision of Hilbert's 1917 course, he expressed himself even more strongly:

As soon as the object of investigation becomes the foundation of...mathematical theories, as soon as one wishes to determine in what relation the theory stands to logic and to what extent it can be obtained from purely logical operations and concepts, then the extended calculus [of logic] is essential.   [Hilbert and Ackermann, 1928, 86]

In the 1917 course (and again in the book), this extended functional calculus permitted quantification over propositions. Hilbert cited the principle of mathematical induction as an axiom that, if fully expressed, requires a quantifier varying over propositions. Likewise, he defined the identity relation in his extended logic in a manner reminiscent of Peirce: two objects x and y are identical if, for every proposition P, P holds of x if and only if P holds of y [Hilbert, 1917, 189–191].

What, in 1917, was Hilbert's extended calculus of logic? At first glance it might appear to have been second-order logic. Yet he stated his preliminary version of the extended calculus so as to permit a function of propositional functions possibly to occupy an argument place for propositions—an act that is illegitimate in second-order logic and that, as he knew, gave rise to Russell's

Paradox. Hilbert used this paradox to motivate his adoption of the ramified theory of types as the definitive version of his extended calculus. He concluded his course by stating: "Thus we have shown that introducing the Axiom of Reducibility is the appropriate means to mold the calculus of orders [the theory of types] into a system in which the foundations of higher mathematics can be developed" [Hilbert, 1917, 246].

In the 1917 course Hilbert treated first-order logic (without any infinitary trappings) as a subsystem of the ramified theory of types:

> In this way is founded a new form of the calculus of logic, the "calculus of orders," which represents an extension of the original functional calculus [first-order logic], since this is contained in it as a theory of first order, but which implies an essential restriction as compared with our previous extension of the functional calculus. [Hilbert, 1917, 222–223]

The same statement appeared in [Hilbert and Ackermann, 1928, 101]. Thus Hilbert regarded first-order logic as a subsystem of all of logic (for him, the ramified theory of types), believing set theory and the principle of mathematical induction to be incapable of adequate treatment in first-order logic [Hilbert, 1917, 189, 200].

At the Zurich lecture, Hilbert invited Paul Bernays to come to Göttingen as his assistant on the foundations of arithmetic. Bernays accepted and soon became Hilbert's principal collaborator in logic. In 1918 Bernays wrote a *Habilitationsschrift* proving the completeness of propositional logic. This work was influenced by Schröder, Frege, and *Principia Mathematica*. For the first time the completeness problem for a subsystem of logic was expressed precisely. Bernays stated and proved that "every provable formula is a valid formula, and conversely" [Bernays, 1918, 6]. In addition he established the deductive completeness of propositional logic: no unprovable formula can be added to the axioms of propositional logic without leading to a contradiction.

Hilbert published nothing further on logic until 1922, when he reacted against L. E. J. Brouwer and Hermann Weyl, who held that the foundations of analysis are built on sand. As a countermeasure, Hilbert introduced his proof theory. This theory treated axiom systems as pure syntax, distinguishing them from what he called metamathematics, where meaning was permitted. His two aims were to show the consistency of both analysis and set theory and to establish the decidability of each mathematical question—aims already expressed in his Paris address [Hilbert, 1900a].

To defend mathematics against Brouwer and Weyl, Hilbert intended to establish the consistency of mathematics from the bottom up. Having shown that a subsystem of arithmetic was consistent, he hoped soon to prove the consistency of the whole of arithmetic (including mathematical induction) and then of the theory of real numbers. Thus Hilbert was thinking in terms of a sequence of levels to be secured successively. In a handwritten appendix to the lecture course, "Logical Foundations of Mathematics," that he gave during 1922–1923, he listed eight of these levels. Analysis occurred at level four, while higher-order logic was considered, in part, at level five.

Already Hilbert [1922, 157] had spoken of the need to formulate Zermelo's Axiom of Choice in such a way that it becomes as evident as $2 + 2 = 4$. In [1923] Hilbert utilized a form of the Axiom of Choice as the cornerstone of his proof theory, which was to be "finitary." He did so as a way of eliminating the direct use of quantifiers, which he regarded as an essentially infinitary feature of logic: "Now where does there appear for the first time something going beyond the concretely intuitive and the finitary? Obviously already in the use of the concepts '*all*' and '*there exists*'" [Hilbert, 1923, 154]. For finite sets, he noted, the universal and existential quantifiers reduce to finite conjunctions and disjunctions, yielding the Principle of the Excluded Middle in the form that $\sim (\forall x) A(x)$ is equivalent to $(\exists x) \sim A(x)$ and that $\sim (\exists x) A(x)$ is equivalent to $(\forall x) \sim A(x)$. "These equivalences," he continued,

> are commonly assumed, without further ado, to be valid in mathematics for infinitely many individuals as well. In this way, however, we abandon the ground of the finitary and enter the domain of transfinite inferences. If we were always to use for infinite sets a procedure admissible for finite sets, we would open the gates to error.... In analysis...the theorems valid for finite sums and products can be translated into theorems valid for infinite sums and products only if the inference is secured, in the particular case, by convergence. Likewise, we must not treat the infinite logical sums and products
>
> $$A(1) \ \& \ A(2) \ \& \ A(3) \ \& \ldots$$

and

$$A(1) \lor A(2) \lor A(3) \lor \ldots$$

in the same way as finite ones. My proof theory...provides such a treatment. [Hilbert, 1923, 155].

Hilbert then introduced what he called the Transfinite Axiom, which he regarded as a form of the Axiom of Choice:

$$A(\tau A) \to A(x).$$

The intended meaning was that if the proposition $A(x)$ holds when x is τA, then $A(x)$ holds for an arbitrary x, say a. He thus defined $(\forall x)A(x)$ as $A(\tau A)$, and similarly for $(\exists x)A(x)$ [Hilbert, 1923, 157]. Soon Hilbert modified the Transfinite Axiom, changing it into the ϵ-axiom:

$$A(x) \to A\big(\epsilon_x(A(x))\big),$$

where ϵ was a choice function acting on properties.

From 1917 to about 1928 Hilbert worked in a version of the ramified theory of types, one in which functions increasingly played the principal role. But by 1928 he had rejected Russell's Axiom of Reducibility as dubious—a significant change from his explicit support for this axiom in [Hilbert, 1917]. Instead, he asserted that the same purpose was served by how he dealt with function variables [Hilbert, 1928, 77]. He still claimed (in [Hilbert and Ackermann, 1928, 114–115]) that the theory of types was the appropriate logical vehicle for studying the theory of real numbers. But, he added, logic could be founded so as to be free of the difficulties posed by the Axiom of Reducibility, as he had done in his various papers. Thus Hilbert opted for a version of the simple theory of types, in effect, ω-order logic.

In [1929] Hilbert looked back with pride, and forward with hope, at what had been accomplished in proof theory. He posed four problems of importance to his program. The first of these was essentially second-order: to prove the consistency of the ϵ-axiom acting on number-theoretic functions. The second was the same problem but stated for higher-order functions. The fourth asked for a demonstration that, on the one hand, the axioms of number theory are deductively complete and that, on the other, first-order logic is complete [1929, 4–8].

Soon Gödel solved this latter problem in his doctoral thesis [1929] by showing the completeness of first-order logic. His abstract [Gödel, 1930a] of this result spoke of the "restricted functional calculus" (first-order logic) as a subsystem of logic, since no bound function variables were permitted.

Although this Completeness Theorem settled one of Hilbert's problems, Gödel [1930b] published a second abstract that threatened to demolish Hilbert's program. For the abstract included the First Incompleteness Theorem: in the theory of types the Peano Postulates are not deductively complete. Gödel's Second Incompleteness Theorem implied that number theory cannot be proved consistent by means of number theory alone but only by a stronger theory. This seemed to destroy any hope of proving the consistency of set theory and analysis with Hilbert's finitary methods.

Probably motivated by Gödel's incompleteness results, Hilbert [1931] introduced a version of the ω-rule, an infinitary rule of inference. In his last statement on proof theory in 1934, Hilbert attempted to limit the damage done by those results. He wrote of

> the final goal of knowing that our customary methods in mathematics are utterly consistent.
>
> Concerning this goal, I would like to stress that the view temporarily widespread—that certain recent results of Gödel imply that my proof theory is not feasible—has turned out to be erroneous. In fact those results show only that, in order to obtain an adequate proof of consistency, one must use the finitary standpoint in a sharper way than is necessary in treating the elementary formalism. (Hilbert in [Hilbert and Bernays, 1934, v])

9. LÖWENHEIM

It was Leopold Löwenheim, working in the Peirce-Schröder tradition, who established an important result about the semantics of logic. His result, the first version of the Löwenheim-Skolem Theorem, is now often considered to be about first-order logic. Yet in one way Löwenheim in 1915 was even farther from first-order logic than Hilbert had been in 1904.

In an autobiographical note, Löwenheim remarked that soon after he began teaching at a Gymnasium (about 1900) he "became acquainted with the calculus of logic through reviews and from Schröder's books" [Thiel, 1977, 237]. Löwenheim's first published paper [1908] was devoted to the solvability of certain symmetric equations in Schröder's calculus of classes, and he soon turned to related questions in Schröder's calculus of domains.

In 1915 there appeared Löwenheim's most influential article, "On Possibilities in the Calculus of Relations." His later opinion of this article was ambivalent:

> I ... have pointed out new paths for science in the field of the calculus of logic, which had been founded by Leibniz but which had come to a deadlock... . On an outing, a somewhat grotesque landscape stimulated my fantasy, and I had an insight that the thoughts I had already developed in the calculus of domains might lead me to make a breakthrough in the calculus of relations. Now I could find no rest until the idea was completely proved, and this gave me a host of troubles... . This breakthrough has scarcely been noticed but some other breakthroughs which I made in my paper "On Possibilities in the Calculus of Relations" were noticed all the same. This became the foundation of the modern calculus of relations. But it was in this very paper that I did not take much pride, since the point had only been to ask the right questions while the proofs could be found easily without ingenuity or imagination.   (Löwenheim in [Thiel, 1977, 246–247])

Löwenheim continued Schröder's work on the logic of relations, making a number of distinctions that Schröder lacked. The most important of these was between a *Relativausdruck* (relational expression) and a *Zählausdruck* (individual expression)[Löwenheim, 1915, 447–448]. Löwenheim's notions of individual expression and relational expression differed from first-order and second-order formulas, respectively, in that, following Peirce and Schröder, these expressions were allowed to have a quantifier for each individual of the domain, or for each relation over the domain. In effect, Löwenheim's logic permitted infinitely long strings of quantifiers and Boolean connectives, but these infinitary formulas occurred only as expansions of finite formulas, to which they were equivalent in his system.

Schröder [1895] had used a logic that was essentially the same as Löwenheim's, namely, a logic that was second order but permitted quantifiers to be expanded as infinitary conjunctions or disjunctions. After distinguishing carefully between the (infinitary) first-order part of his logic and the entire logic, Löwenheim asserted that all important problems of mathematics can be handled in this (infinitary) second-order logic.

Apparently, Löwenheim's Theorem was motivated by the two ways in which Schröder's calculus of relations could express propositions: (1) by means of quantifiers and individual variables and (2) by means of relations with no individual variables. Whereas Schröder regarded every proposition of form (1) as capable of being "condensed" (that is, written in form (2)) because he permitted individuals to be interpreted as a certain kind of relation, Löwenheim dispensed with this interpretation and then showed that not all propositions can be condensed. The reason was that condensable first-order propositions could not express that there are at least four elements.

After treating condensation, Löwenheim immediately stated an analogous result for denumerable domains: Given a domain M that is infinite (possibly denumerable and possibly of higher cardinality), if a first-order proposition is valid in every finite domain but not in every infinite domain, then the proposition is not valid in M. This was his original version of Löwenheim's Theorem. His proof used a second-order formula asserting that

$$(\forall x)(\exists y)\phi(x,y) \leftrightarrow (\exists f)(\forall x)\phi(x,f(x)),$$

where f was a function variable, and he regarded this formula (for the given domain M) as expandable to an infinitary formula with a first-order existential quantifier for each individual of M [Löwenheim, 1915, Section 2]. His next theorem stated that, since Schröder's logic can express that a domain is finite or denumerable, then Löwenheim's Theorem cannot be extended to Schröder's (second-order) logic.

Some historians have regarded Löwenheim's argument establishing Löwenheim's Theorem as odd and unnatural. But this argument appears so only because they have rephrased it within first-order logic. While the theorem can be shown for first-order logic (as Skolem was soon to do), this was not the logic that Löwenheim used.

10.   SKOLEM: FIRST-ORDER LOGIC AS ALL OF LOGIC

In 1913, after writing a thesis on Schröder's algebra of logic, Thoralf Skolem received his undergraduate degree in mathematics. Subsequently, he wrote several papers dealing with Schröder's calculus of classes. During the winter of 1915-1916 Skolem visited Göttingen, where he discussed set theory with Felix Bernstein. By that time Skolem already knew of Löwenheim's Theorem and had seen how to extend it to any countable set of formulas. Moreover, Skolem had applied this extended version of the theorem to set theory, thereby obtaining what was later called *Skolem's Paradox*: Set theory has a countable model (within first-order logic) even though set theory yields the existence of uncountable sets [Skolem, 1923, 232, 219].

Nevertheless, Skolem did not lecture on this result until 1922, and it did not appear in print until 1923. When the result appeared, he remarked that he had not published it earlier because he had been occupied with other problems and also because

> I believed that it was so clear that the axiomatization of set theory would not be satisfactory as an ultimate foundation for mathematics that, by and large, mathematicians would not bother themselves with it very much. To my astonishment I have seen recently that many mathematicians regard these axioms for set theory as the ideal foundation for mathematics. For this reason it seemed to me that the time had come to publish a critique. [Skolem, 1923, 232]

It was in order to establish the relativism of set-theoretic notions that Skolem suggested that set theory should be formulated within first-order logic. At first glance, given the historical context, this was a strange suggestion; for set theory seemed to require quantifiers not only over individuals, as in first-order logic, but also quantifiers over sets of individuals, over sets of sets of individuals, and so forth. Skolem's radical suggestion was that the membership relation ϵ be treated not as a part of logic (as previous logicians had done) but simply as any other relation that could be given a variety of interpretations in various domains.

In [1920] Skolem began to publish his research on Löwenheim's Theorem—a subject to which he returned repeatedly over the next 40 years. First, Skolem gave a new proof of Löwenheim's Theorem that avoided using second-order logic. Then he extended this result to the Löwenheim-Skolem Theorem for first-order logic (if a countable set of first-order propositions has a model, then it has a countable model), but he did not state it as a separate theorem. Instead, he proved Löwenheim's Theorem for countably infinite conjunctions and disjunctions of first-order formulas. Yet when he applied Löwenheim's Theorem to set theory in 1923, Skolem stated the Löwenheim-Skolem Theorem in the usual form. Henceforth he never went back to the infinitary logic that he had inherited from Schröder and Löwenheim.

The Löwenheim-Skolem Theorem, especially as applied to set theory, received mixed reviews. Abraham Fraenkel, who reviewed [Skolem, 1923] in the abstracting journal *Jahrbuch über die Fortschritte der Mathematik*, did not mention first-order logic. Fraenkel seemed to think that, thanks to Skolem's results, the relativism of the notion of cardinal number was inherent in every axiomatic system. Thus Fraenkel did not understand the divergent effects of first-order and second-order logic on the notion of infinite cardinal number.

In 1925, when John von Neumann published his axiomatization for set theory, he too was unclear about the difference between first-order and second-order logic. He specified his desire to axiomatize set theory by means of "a finite number of purely formal operations" [von Neumann, 1925, Section 2], but nowhere did he specify the logic in which his axiomatization was to be formulated. His concern, rather, was to give a finite characterization of the notion of "definite property" that Zermelo had used in axiomatizing set theory. Von Neumann considered the question of categoricity in detail and noted various steps, such as elimination of inaccessible cardinals, needed to obtain a categorical axiomatization for set theory. After remarking that the axioms for Euclidean geometry were categorical, he observed that, as a result of the

Löwenheim-Skolem Theorem,

> no categorical axiomatization of set theory seems to exist at all ... . And
> since there is no axiom system for mathematics, geometry, and so forth that
> does not presuppose set theory, there probably cannot be any categorically
> axiomatized infinite systems at all. This circumstance seems to me to be an
> argument for intuitionism.   [von Neumann, 1925, Section 5]

Then von Neumann argued that the boundary between the finite and infinite was also blurred. He did not realize that none of these difficulties occurred in second-order logic. For at the time the distinction between first-order logic and second-order logic remained unclear, and it was equally unclear just how widely the Löwenheim-Skolem Theorem applied.

In [1929] Ernst Zermelo responded to the criticisms—by Fraenkel, Skolem, and von Neumann—of his notion of "definite property." Zermelo, in effect, chose to define a "definite property" as any second-order formula built up from the primitive symbols $=$ and ϵ, rather than as a first-order formula built up the same way (as Skolem [1923] had done). Zermelo [1930] used second-order logic to prove that any standard model of Zermelo-Fraenkel set theory consists of V_α for some level of Zermelo's cumulative type hierarchy if and only if α is an inaccessible ordinal—a result that holds in second-order logic but fails in first-order logic.

Zermelo was dismayed by the uncritical acceptance of what he called "*Skolemism*, the doctrine that *every* mathematical theory, and set theory in particular, is satisfiable in a *countable model*" [Zermelo, 1931, 85]. The logic that Zermelo proposed was an infinitary logic with conjunctions and disjunctions of any infinite cardinality, but no quantifiers. It was by far the strongest infinitary logic proposed up to that time.

Zermelo, unhappy too with Gödel's Incompleteness Theorem, wrote to him for clarification. There followed a spirited exchange of letters between the old set-theorist and the young logician. Zermelo's letters reveal that, like Skolem and many others, he did not distinguish clearly between syntax and semantics. On the other hand, Zermelo insisted on extending the notion of proof far enough so that every valid formula would be provable. Zermelo, in his last published paper [1935], extended the notion of rule of inference to allow not only the ω-rule, which Hilbert had introduced in 1931, but arbitrarily long well-founded strings of premises.

Yet Zermelo's ideas about extending logic met with little sympathy. Gödel, who had adopted Zermelo's cumulative type hierarchy, formulated set theory within first-order logic. Indeed, Gödel's constructible sets (which enabled him to prove the relative consistency of the Axiom of Choice and the Generalized Continuum Hypothesis) were the natural fusion of first-order logic with the cumulative type hierarchy.

In December 1938, at a conference in Zurich on the foundations of mathematics, Skolem returned again to the existence of countable models for set theory and to the Skolem Paradox [Skolem, 1941]. But on this occasion he chose to emphasize the relativism, not only of set theory, but of mathematics as a whole. The discussion that followed Skolem's lecture revealed both interest and ambivalence about his result. In particular, Bernays commented at length:

> The axiomatic restriction of the notion of set [to first-order logic] does not prevent one from obtaining all the usual theorems...of Cantorian set theory.... Nevertheless,...this way of making the notion of set (or that of predicate) precise has a consequence of another kind: the interpretation of the system is no longer necessarily unique.... The impossibility of characterizing the finite with respect to the infinite comes from the restrictiveness of the [first-order] formalism. The impossibility of characterizing the denumerable with respect to the nondenumerable in a sense that is in some way unconditional—does this reveal, one might ask, a certain inadequacy of the method under discussion here [first-order logic] for making axiomatizations precise?   (Bernays in [Gonseth, 1941, 49–50])

Skolem objected vigorously to Bernays' suggestion, and argued that a first-order axiomatization is indeed the most appropriate.

11. CONCLUSION

To sum up, the logics introduced from 1879 to 1923 were richer than first-order logic. This richness took two forms: the use of infinitely long expressions (by Peirce, Schröder, Hilbert, Löwenheim, and Skolem) and the use of a logic at least as rich as second-order logic (by Frege, Peirce, Schröder, Löwenheim, Peano, Russell, and Hilbert). The fact that no system of logic predominated (although the Peirce-Schröder tradition was strong until about 1920, as was *Principia Mathematica* during the 1920s and 1930s) encouraged both variety and richness in logic.

First-order logic emerged as a distinct *subsystem* of logic in Hilbert's lectures of 1917 and, in print, in [Hilbert and Ackermann, 1928]. Nevertheless, Hilbert did not regard first-order logic as the proper basis for mathematics. From 1917 on, he worked in the theory of types.

It is in Skolem's work on set theory [Skolem, 1923] that first-order logic was first proposed as all of logic and that set theory was first formulated within first-order logic. For the next four decades Skolem tried to convince the mathematical community that both of his proposals were correct. The first claim, that first-order logic is all of logic, was taken up by W. V. Quine, who argued that second-order logic is really set theory in disguise [Quine, 1941, 144–145]. This claim fared well for a while. After a distinct infinitary logic emerged in the 1950s (thanks to Alfred Tarski) and after the introduction of generalized quantifiers (thanks to Andrzej Mostowski [1957]), first-order logic was clearly not all of logic. Skolem's second claim, that set theory should be formulated in first-order logic, was quite successful, and today this is how almost all set theory is done.

When Gödel [1929; 1930] proved the completeness of first-order logic and then the incompleteness of both second-order and ω-order logic [Gödel, 1931], he both stimulated first-order logic and inhibited the growth of second-order logic. Yet his incompleteness results encouraged the search for an appropriate infinitary logic—by Hilbert [1931], Rudolf Carnap [1935], and Zermelo [1935]. During the 1930s and 1940s first-order logic was gradually accepted as the basis of all of mathematics.

Yet Maltsev [1936], through the use of uncountably many symbols in first-order logic, and Tarski, through the Upward Löwenheim-Skolem Theorem, rejected Skolem's attempt to restrict logic to countable first-order languages. Uncountable first-order languages and uncountable models have become a standard part of first-order logic. Thus set theory entered logic through the back door, both syntactically and semantically, though it failed to enter through the front door of second-order logic.

NOTES

[1] Since this paper is intended for nonlogicians, a glossary of logical terms is included after these notes. Mathematical logicians are referred to a different version, entitled "The Emergence of First-Order Logic," to appear in 1987 in a volume *History and Philosophy of Modern Mathematics* (of *Minnesota Studies in the Philosophy of Science*) edited by P. Kitcher and W. Aspray.

[2] Peano acknowledged [Peano, 1891, 93] that his postulates for the natural numbers came from [Dedekind, 1888].

[3] There are only countably many such definable subsets because there are only countably many definitions, each a finite string of symbols.

[4] To give a nonmathematical example, in the sentence "Patience is a virtue" patience may be regarded as a property of individuals (in this example, human beings) while virtue is a property possessed by patience and hence is a second-order concept (or property) in Frege's sense.

[5] [Schröder, 1895, 514]. As an example, let the domain consist of the natural numbers and consider the (false) proposition that for each relation there exist numbers x and y in the relation. For Schröder, this proposition could be expanded to a logically equivalent infinitary proposition. If $R_1, R_2, \ldots$ are all the relations on the natural numbers, then the expanded infinitary proposition is the following: There exist $x_1, x_2, \ldots$ such that x_1 is in the relation R_1 to x_2, and x_3 is in the relation R_2 to x_4, and $\ldots$.

[6] Frege pointed this out to Hilbert in a letter of January 6, 1900; see [Frege, 1980, 46]. Frege discussed the matter in print in [Frege, 1903a, 370–371].

[7] [Frege, 1980, 108]. Peano [1897] introduced a separate symbol for the existential quantifier.

GLOSSARY

Axiom of Reducibility: an axiom stating, in the ramified theory of types, that every proposition is equivalent to a first-order proposition.

Completeness Theorem (for first-order logic):   A proposition is provable (in an axiom system) if and only if it is true in every model (of the system).

deductive completeness: An axiom system is deductively complete if every proposition expressible in the system is either provable or disprovable in the system.

Habilitationsschrift: a higher doctorate, required in order to give lecture courses at a German (or Swiss) university.

inaccessible cardinal: a cardinal that is not the sum or product of a smaller number of smaller cardinals.

Incompleteness Theorems: The First Incompleteness Theorem states that any axiom system rich enough to express number theory is not deductively complete (and there is a proposition that is true in the system but not provable in it). The Second Incompleteness Theorem states that, in any axiom system S rich enough to express number theory, one cannot prove the consistency of S.

Löwenheim's Theorem: If a proposition is true in every finite set but not in every set, then there is a denumerable set in which the proposition is not true.

Löwenheim-Skolem Theorem: In first-order logic, if a set of propositions has a model, it has a countable model.

metalanguage: the language in which one is speaking, as opposed to the language that one is speaking about.

ω-rule: For a proposition $A(x)$: From $A(0)$, $A(1)$, $A(2)\ldots$, infer $(\forall x)A(x)$.

Principle of Comprehension: the postulate that to every formula with one free variable there is a set consisting of those objects satisfying the formula.

Principle of the Excluded Middle: the proposition "A or not A".

Russell's Paradox: Let S be the set of all sets x such that x is not a member of x. Then S is a member of S if and only if S is not a member of S.

Upward Löwenheim-Skolem Theorem: In first-order logic, if a set of propositions has a model of some infinite cardinality, it has a model of every larger cardinality.

FURTHER READING

For a general introduction to late nineteenth and early twentieth century logic, see [Kneale and Kneale, 1962, Chapters 6–12]. On the history of the algebra of logic, see [Lewis, 1918]. Concerning the history of the ω-rule, see [Feferman, 1986]. For the relation between set theory and logic, see [Moore, 1980] and [Moore, 1982, 249–283]. On semantics from Löwenheim to the Second World War, see [Vaught, 1974]. Two source books in English are [Bochenski, 1970] and [van Heijenoort, 1967]. Finally, for a defense of second-order logic, see [Boolos, 1975] and [Shapiro, 1985].

REFERENCES

Bernays, P., 1918, *Beiträge zur axiomatischen Behandlung des Logik-Kalküls*, Habilitationsschrift, Göttingen.

Bochenski, I., 1970, *A History of Formal Logic*, Chelsea, New York.

Boole, G., 1854. *An Investigation of the Laws of Thought on Which Are Founded the Mathematical Theories of Logic and Probabilities*, London.

Boolos, G., 1975, "On Second-Order Logic," *Journal of Philosophy* 72, 509–527.

Bynum, T. W., 1972, *Conceptual Notation and Related Articles*, Clarendon Press, Oxford.

Carnap, R., 1935, "Ein Gültigkeitskriterium für die Sätze der klassischen Mathematik," *Monatshefte für Mathematik und Physik* 41, 263–284.

Dedekind, R., 1888, *Was sind und was sollen die Zahlen*? Braunschweig.

De Morgan, A., 1859. "On the Syllogism No. IV, and on the Logic of Relations," *Transactions of the Cambridge Philosophical Society* 10, 331–358.

Feferman, S., 1986, Introductory Note, in [Gödel, 1986, 208–213].

Fisch, M. H., 1984, "The Decisive Year and Its Early Consequences," in [Peirce, 1984, xxi–xxvi].

Frege, G., 1879, *Begriffsschrift, eine der arithmetischen nachgebildete Formelsprache des reinen Denkens*, Louis Nebert, Halle. English translations in [Bynum, 1972] and in [van Heijenoort, 1967].

———, 1883, "Über den Zweck der Begriffsschrift," *Sitzungsberichte der Jenaischen Gesellschaft für Medicin und Naturwissenschaft* 16, 1–10. English translation in [Bynum, 1972, 90–100].

———, 1884, *Grundlagen der Arithmetik*: *Ein logisch-mathematische Untersuchung über den Begriff der Zahl*, Koebner, Breslau.

———, 1893, *Grundgesetze der Arithmetik, begriffsschriftlich abgeleitet*, Volume 1, Pohle, Jena.

———, 1903, Volume 2 of [1893].

———, 1903a, "Über die Grundlagen der Geometrie," *Jahresbericht der Deutschen Mathematiker-Vereinigung* 12, 319–324.

———, 1980, *Philosophical and Mathematical Correspondence*, edited by G. Gabriel, H. Hermes, F. Kambartel, C. Thiel, and A. Veraart. Abridged by B. McGuinness and translated by H. Kaal, University of Chicago Press. Chicago.

Gentzen, G., 1936, "Die Widerspruchsfreiheit der Stufenlogik," *Mathematische Zeitschrift* 41, 357–366.

Gödel, K., 1929, "Über die Vollständigkeit des Logikkalküls," doctoral dissertation: University of Vienna. Printed, together with an English translation, in [Gödel, 1986, 60–101].

_____, 1930, "Die Vollständigkeit der Axiome des logischen Funktionenkalküls," *Monatshefte für Mathematik und Physik* 37, 349–360. Reprinted, with English translation, in [Gödel, 1986, 102–123].

_____, 1930a, "Über die Vollständigkeit des Logikkalküls," *Die Naturwissenschaften* 18, 1068. Reprinted, with English translation, in [Gödel, 1986, 124–125].

_____, 1930b, "Einige metamathematische Resultate über Entscheidungsdefinitheit und Widerspruchsfreiheit," *Anzeiger der Akademie der Wissenschaften in Wien* 67, 214–215. Reprinted, with English translation, in [Gödel, 1986, 140–143].

_____, 1931, "Über formal unentscheidbare Sätze der Principia Mathematica und verwandter Systeme I," *Monatshefte für Mathematik und Physik*, 38, 173–198. Reprinted, with English translation, in [Gödel, 1986, 126–195].

_____, 1986, *Collected Works*, edited by S. Feferman, J. W. Dawson, Jr., S. C. Kleene, G. H. Moore, R. M. Solovay, and J. van Heijenoort, Volume I: *Publications 1929–1936*, Oxford University Press, New York.

Gonseth, F. (ed.), 1941, *Les entretiens de Zurich, 6–9 décembre 1938*, Leeman, Zurich.

Hilbert, D., 1899, *Grundlagen der Geometrie. Festschrift zur Feier der Enthüllung des Gauss-Weber Denkmals in Göttingen*, Teubner, Leipzig.

_____, 1900, "Über den Zahlbegriff," *Jahresbericht der Deutschen Mathematiker-Vereinigung* 8, 180–194.

_____, 1900a, "Mathematische Probleme, Vortrag, gehalten auf dem internationalem Mathematiker-Kongress zu Paris, 1900," *Nachrichten von der Königlichen Gesellschaft der Wissenschaften zu Göttingen*, 253–297.

_____, 1902, *Les principes fondamentaux de la géometrie*, Gauthier-Villars, Paris. French translation of [Hilbert, 1899] by L. Laugel.

_____, 1903, *The Foundations of Geometry*, Open Court, Chicago. English translation of [Hilbert, 1899] by E. J. Townsend.

_____, 1905, "Über die Grundlagen der Logik und der Arithmetik," *Verhandlungen des dritten internationalen Mathematiker-Kongresses in Heidelberg vom 8. bis 13. August 1904*, Teubner, Leipzig, 1905. English translation in [van Heijenoort, 1967, 129–138].

_____, 1917, *Prinzipien der Mathematik und Logik*. Lecture notes of a course given at Göttingen in the winter semester of 1917–1918.

_____, 1918, "Axiomatisches Denken," *Mathematische Annalen* 78, 405–415. Reprinted in [Hilbert, 1935, 178–191].

_____, 1922, "Neubegründung der Mathematik (Erste Mitteilung)," *Abhandlungen aus dem mathematischen Seminar der Hamburgischen Universität* 1, 157–177. Reprinted in [Hilbert, 1935, 157–177].

_____, 1923, "Die logischen Grundlagen der Mathematik," *Mathematische Annalen* 88, 151–165. Reprinted in [Hilbert, 1935, 178–191].

_____, 1926, "Über das Unendliche," *Mathematische Annalen* 95, 161–190. English translation in [van Heijenoort, 1967, 367–392].

_____, 1928, "Die Grundlagen der Mathematik," *Abhandlungen aus dem mathematischen Seminar der Hamburgischen Universität* 6, 65–85. English translation in [van Heijenoort, 1967, 464–479].

_____, 1929, "Probleme der Grundlegung der Mathematik," *Mathematische Annalen* 102, 1–9.

_____, 1931, "Die Grundlegung der elementaren Zahlenlehre," *Mathematische Annalen* 104, 485–494. Reprinted in part in [Hilbert, 1935, 192–195].

_____, 1935, *Gesammelte Abhandlungen*, Volume 3, Springer, Berlin.

Hilbert, D., and W. Ackermann, 1928, *Grundzüge der theoretischen Logik*, Springer, Berlin.

Hilbert, D., and P. Bernays, 1934, *Grundlagen der Mathematik*, Volume 1, Springer, Berlin.

_____, 1939, Volume 2 of [1934].

Huntington, E. V., 1902, "A Complete Set of Postulates for the Theory of Absolute Continuous Magnitude," *Transactions of the American Mathematical Society* 3, 264–279.

_____, 1905, "A Set of Postulates for Ordinary Complex Algebra," *Transactions of the American Mathematical Society* 6, 209–229.

Kneale, W. and M., 1962, *The Development of Logic*, Clarendon Press, Oxford.

Lewis, C. I., 1918, *A Survey of Symbolic Logic*, University of California Press, Berkeley.

Löwenheim, L., 1908, "Über das Auslösungsproblem im logischen Klassenkalkül," *Sitzungsberichte der Berliner Mathematischen Gesellschaft*, 89–94.

_____, 1915, "Über Möglichkeiten im Relativkalkül," *Mathematische Annalen* 76, 447–470. English translation in [van Heijenoort, 1967, 228–251].

Maltsev, A. I., 1936, "Untersuchungen aus dem Gebiete der mathematischen Logik," *Matematicheskii Sbornik* 1, 323–336.

Moore, G. H., 1980, "Beyond First-Order Logic: The Historical Interplay between Mathematical Logic and Axiomatic Set Theory," *History and Philosophy of Logic* 1, 95–137.

_____, 1982, *Zermelo's Axiom of Choice: Its Origins, Development, and Influence*, Springer, New York.

Peano, G., 1888, *Calcolo geometrico secondo l'Ausdehnungslehre di H. Grassmann, preceduto dalle operazioni della logic deduttiva*, Bocca, Turin.

_____, 1889, *Arithmetices principia, nova methodo exposita*, Bocca, Turin. Partial English translation in [van Heijenoort, 1967, 83–97].

_____, 1891, "Sul concetto di numero," *Rivista di Matematica* 1, 87–102, 256–267.

_____, 1897, "Studii di logica matematica," *Atti della Reale Accademia delle Scienze di Torino* 32, 565–583.

Peirce, C. S., 1865, "On an Improvement in Boole's Calculus of Logic," *Proceedings of the American Academy of Arts and Sciences* 7, 250–261.

_____, 1870, "Description of a Notation for the Logic of Relatives, Resulting from an Amplification of the Conceptions of Boole's Calculus of Logic," *Memoirs of the American Academy* 9, 317–378.

_____, 1883, "The Logic of Relatives," in *Studies in Logic*, edited by C. S. Peirce, Little, Brown, and Company, Boston, pp. 187–203.

_____, 1885, "On the Algebra of Logic: A Contribution to the Philosophy of Notation," *American Journal of Mathematics* 7, 180–202.

———, 1976, *The New Elements of Mathematics*, edited by C. Eisele, Volume III: *Mathematical Miscellanea*, Mouton, Paris.

———, 1984, *Writings of Charles S. Peirce, A Chronological Edition*, edited by E. C. Moore, Volume 2, 1867–1871, Indiana University Press, Bloomington.

Poincaré, H., 1905, "Les mathématiques et la logique," *Revue de Métaphysique et de morale* 13, 815–835.

Quine, W. V., 1941, "Whitehead and the Rise of Modern Logic," in *The Philosophy of Alfred North Whitehead*, edited by P. A. Schilpp (Northwestern University, Evanston), pp. 125–163.

Russell, B., 1903, *The Principles of Mathematics*, Cambridge University Press, Cambridge.

———, 1908, "Mathematical Logic as Based on the Theory of Types," *American Journal of Mathematics* 30, 222–262.

Schröder, E., 1877, *Der Operationskreis des Logikkalküls*, Teubner, Leipzig.

———, 1880, review of [Frege, 1879], *Zeitschrift für Mathematik und Physik* 25, 81–94. English translation in [Bynum, 1972, 218–232].

———, 1890, *Vorlesungen über die Algebra der Logik (exacte Logik)*, Volume 1, Leipzig.

———, 1891, Volume 2, part 1, of [Schröder, 1890].

———, 1895, Volume 3 of [Schröder, 1890].

Shapiro, S., 1985, "Second-Order Languages and Mathematical Practice," *Journal of Symbolic Logic*, 50, 714–742.

Skolem, T., 1913, *Undersøkelser innenfor logikkens algebra (Researches on the Algebra of Logic)*, undergraduate thesis, University of Oslo.

———, 1920, "Logisch-kombinatorische Untersuchungen über die Erfüllbarkeit oder Beweisbarkeit mathematischer Sätze nebst einem Theoreme über dichte Mengen," *Videnskapsselskapets skrifter, I. Matematisk-naturvidenskabelig klasse*, no. 4. English translation of Section 1 in [van Heijenoort, 1967, 252–263].

———, 1923, "Einige Bemerkungen zur axiomatischen Begründung der Mengenlehre," *Videnskapsselskapets skrifter, I. Matematisk-naturvidenskabelig klasse*, no. 6. English translation in [van Heijenoort, 1967, 302–333].

———, 1933, "Über die Möglichkeit einer vollständigen Charakterisierung der Zahlenreihe mittels eines endlichen Axiomensystems," *Norsk matematisk forenings skrifter*, series 2, no. 10, 73–82. Reprinted in [Skolem, 1970, 345–354].

———, 1934, "Über die Nicht-charakterisierbarkeit der Zahlenreihe mittels endlich oder abzählbar unendlich vieler Aussagen mit ausschliesslich Zahlenvariablen," *Fundamenta Mathematicae* 23, 150–161. Reprinted in [Skolem, 1970, 355–366].

———, 1941, "Sur la portée du théorème de Löwenheim-Skolem," in [Gonseth, 1941, 25–52]. Reprinted in [Skolem, 1970, 455–482].

———, 1970, *Selected Works in Logic*, edited by J. E. Fenstad, Universitetsforlaget, Oslo.

Thiel, C., 1977, "Leopold Löwenheim: Life, Work, and Early Influence," in *Logic Colloquium 76*, edited by R. Gandy and M. Hyland, North-Holland, Amsterdam, pp. 235–252.

van Heijenoort, J., 1967, *From Frege to Gödel: A Source Book in Mathematical Logic*, Harvard University Press, Cambridge.

Vaught, R. L., 1974, "Model Theory before 1945," in *Proceedings of the Tarski Symposium*, edited by L. Henkin, Volume 25 of *Proceedings of Symposia in Pure Mathematics*, American Mathematical Society, New York, pp. 153–172.

Veblen, O., 1904, "A System of Axioms for Geometry," *Transactions of the American Mathematical Society* 5, 343–384.

von Neumann, J., 1925, "Eine Axiomatisierung der Mengenlehre," *Journal für die reine und angewandte Mathematik* 154, 219–240. English translation in [van Heijenoort, 1967, 393–413].

Whitehead, A. N. and B. Russell, 1910, *Principia Mathematica*, vol. 1, Cambridge University Press, Cambridge.

Zermelo, E., 1929, "Über den Begriff der Definitheit in der Axiomatik," *Fundamenta Mathematicae* 14, 339–344.

_____, 1930, "Über Grenzzahlen und Mengenbereiche: Neue Untersuchungen über die Grundlagen der Mengenlehre," *Fundamenta Mathematicae* 16, 29–47.

_____, 1931, "Über Stufen der Quantifikation und die Logik des Unendlichen," *Jahresbericht der Deutschen Mathematiker-Vereinigung* 41, Angelegenheiten, 85–88.

_____, 1935, "Grundlagen einer allgemeinen Theorie der mathematischen Satzsysteme," *Fundamenta Mathematicae* 25, 136–146.

MATHEMATICAL LOGIC AND THE ORIGIN OF MODERN COMPUTERS

Martin Davis

The very word *computer* immediately suggests one of the main uses of these remarkable devices: an instrument of calculation. But it is a matter of widespread experience that modern computers can be used for many purposes having no evident connection with numerical computation. The main thesis of this article is that the source of this generalized conception of the scope of computers is to be found in the vision of a computer as an engine of logic implicit in the abstract theory of computation developed by mathematical logicians.

The connection between logic and computing is apparent even from the everyday use of language: the English word "reckon" means both to calculate and to conclude. Without trying to understand this connection in any very profound manner, we can certainly see that computation is a (very restricted) form of reasoning. To see the connection in the opposite direction, imagine our seeking to demonstrate to a skeptic that some conclusion follows logically from certain assumptions. We present a "proof" that our claim is correct, only to be faced by the demand that we demonstrate that our proof is correct. If we then attempt a proof that our previous "proof" was correct, we clearly are faced with an infinite regress. The way out that has been found is to insist on a purely algorithmic criterion for logical correctness—a proof is correct if it proceeds according to rules whose correct application can be verified in a purely computational manner.

There are many examples of important concepts and methods first introduced by logicians which later proved to be important in computer science. Tracing the paths along which some of these ideas found their way from theory to practice is a fascinating (and often frustrating) task for the historian of ideas. The subject of this paper is Alan Turing's discovery of the universal (or all-purpose) digital computer as a mathematical abstraction. This concept was introduced by Turing as part of his solution to a problem that David Hilbert had called the "principal problem of mathematical logic" [Hilbert and Ackermann, 1928]. We will try to show how this very abstract work helped to lead Turing and John von Neumann to the modern concept of the electronic computer.

But first, before discussing the work of Alan Turing, we will see how some of the underlying themes of computer science had already appeared in the seventeenth century in the work of G. W. Leibniz (1646–1716).

LEIBNIZ' DREAM

It is striking to note the many different ways in which Gottfried Leibniz anticipated what later came to be central concerns in computer science. He made an important invention, the so-called *Leibniz wheel*, which he used as early as the 1670s to build a mechanical calculating machine that could add, subtract, multiply, and divide. He showed keen awareness of the great advantages to be expected from the mechanization of computation. Thus, Leibniz [Smith, 1929, 180–181] said of his calculator:

> And now that we may give final praise to the machine we may say that it will be desirable to all who are engaged in computations...managers of financial affairs, merchants, surveyors, geographers, navigators, astronomers.... But limiting ourselves to scientific uses, the old geometric and astronomic tables could be corrected and new ones constructed.... it will pay to extend as far as possible the major Pythagorean tables; the table of squares, cubes, and

other powers; and the tables of combinations, variations, and progressions of all kinds,...Also the astronomers surely will not have to continue to exercise the patience which is required for computation....For it is unworthy of excellent men to lose hours like slaves in the labor of computation.

Leibniz was one of the first [Ceruzzi, 1983, 40, footnote 11] to work out the properties of the binary number system, which of course has turned out to be fundamental for computer science. He proposed the development of a *calculus of reason* or *calculus ratiocinator* and actually proceeded to develop what amounts to a fragment of Boolean algebra ([Parkinson, 1966, 132–133], [Davis, 1983, 2–3]). Finally, there was Leibniz' amazing program calling for the development of a universal language—a *lingua characteristica*—which would not only incorporate the calculus ratiocinator, but would also be suitable for communication and would include scientific and mathematical knowledge. Leibniz hoped to mechanize much of thought, saying that the mind "will be freed from having to think directly of things themselves, and yet everything will come out correctly" [Parkinson, 1966, xvii]. Leibniz imagined problems in human affairs being handled by a learned committee sitting around a table and saying [Kneale and Kneale, 1962, 328], "Lasst uns rechnen!" i.e., "Let us calculate!"

The importance with which Leibniz regarded these projects is clear from his assessment:

> For if praise is given to the men who have determined the number of regular solids—which is of no use, except insofar as it is pleasant to contemplate—and if it is thought to be an exercise worthy of a mathematical genius to have brought to light the more elegant properties of a conchoid or cissoid, or some other figure which rarely has any use, how much better will it be to bring under mathematical laws human reasoning, which is the most excellent and useful thing we have.   [Parkinson, 1966, 105]

It is at times amusing to imagine some great person from a past age reacting to one of the marvels of the contemporary world. Confronted by a modern computer, Leibniz would surely have been awestruck by the wonders of twentieth century technology. But perhaps he would have been better equipped than any other seventeenth century person to comprehend the scope and potential of these amazing machines.

ALAN TURING'S ANALYSIS OF THE CONCEPT OF COMPUTATION

A century and a half ago, Charles Babbage already had conceived of an all-purpose automatic calculating machine, his proposed but never constructed *analytical engine*. Babbage's device
was intended to carry out computations of the most varied kind
that arise in algebra and mathematical analysis. To emphasize the
power and scope of his engine, Babbage remarked facetiously that
"it could do everything but compose country dances" [Huskey and
Huskey, 1980, 300]. A contemporary computer expert seeking a
figure of speech to bring home to a popular audience the widespread
applicability of computers would select a different example—for
we know that today's computers can perfectly well be programmed
to compose country dances (although presumably not of the finest
quality). While for Babbage it was self-evident that calculating
machines could not be expected to compose dances, it does not
strike us today as being at all out of the question. Clearly, our very
concept of what constitutes "computation" has been altered
drastically. We shall see how this expanded concept of computation developed out of the work in mathematical logic of Alan
Turing.

Babbage never succeeded in constructing his engine, in large part
because of the limitations of nineteenth century technology. In
fact, it was only with some of the electro-mechanical calculators
that began to be built during the 1930s (for example, by Howard
Aiken at Harvard University) that Babbage's vision was fully
realized. But during the 1930s and 1940s no one involved with this
work suggested the possibility of designing an automatic computer
that not only could do everything that Babbage had envisioned, but
also could be used for commercial purposes, or for that matter, to
"compose country dances." Even as late as 1956, Howard Aiken,
himself a pioneer of modern computing, could write:

> If it should turn out that the basic logics of a machine designed for the
> numerical solution of differential equations coincide with the logics of a
> machine intended to make bills for a department store, I would regard this as
> the most amazing coincidence that I have ever encountered. [Ceruzzi, 1983,
> 43]

If Aiken had grasped the significance of a paper by Alan Turing that had been published two decades earlier [Turing, 1936], he would never have found himself in the position of making such a statement only a few years before machines that performed quite well at both of the tasks he listed were readily available.

Alan Mathison Turing was born on June 23, 1912 in London. His father was a civil servant in India, and Turing spent most of his childhood away from his parents.[1] After five years at Sherborne, a traditional English public school, he was awarded a fellowship to King's College at Cambridge University. Turing arrived at Cambridge in 1931. This was just after the young logician Kurt Gödel had startled the mathematical world by demonstrating that for any formal system adequate for elementary number theory, arithmetic assertions could be found that were not decidable within that formal system. In fact, Gödel had even shown that among these "undecidable propositions" was the very assertion that the given formal system itself is consistent. This last result was devastating to Hilbert's program in the foundations of mathematics, which called for proving the consistency of more and more powerful formal systems using only very restricted proof methods, methods that Hilbert called *finitistic*. John von Neumann was probably the most brilliant of the young people who had been striving to carry out Hilbert's program. In addition to his contributions to Hilbert's program, von Neumann's intense interest in logic and foundations is also evidenced by his early papers on axiomatic set theory [von Neumann, 1961]. However, after Gödel's discovery, von Neumann stopped working in this field. In the spring of 1935, Turing attended a course of lectures by the topologist M. H. A. Newman, on *Foundations of Mathematics* in which Hilbert's program and Gödel's work were among the topics discussed. In particular, Newman called the attention of his audience to Hilbert's *Entscheidungsproblem*, a problem which Hilbert had called the "principal problem of mathematical logic."

In 1928, a little textbook of logic by Hilbert and Wilhelm Ackermann, entitled *Grundzüge der theoretischen Logik*, had been published. The book emphasized first-order logic, the logic of *and*, *or*, *not*, *if … then*, *for all*, and *there exists*, which the authors called the *engere Funktionenkalkül*. The authors showed how the various parts of mathematics could be formalized within first-order logic, and a simple set of rules of proof were given for making logical inferences. They noted that any inference that can be carried out according to their rules of proof is also *valid*, in the sense that in any mathematical structure in which all the premises are true, the conclusion is also true. Hilbert and Ackermann then raised the problem of *completeness*: if an inference is valid (in the sense just explained), would it always be possible, using their rules of proof, to obtain the conclusion from the premises? This question was answered affirmatively two years later by Gödel in his doctoral dissertation at the University of Vienna. Another problem raised in the *Gründzuge* by Hilbert and Ackermann was the *Entscheidungsproblem*, the problem of finding an algorithm to determine whether a given proposed inference is valid. By the completeness theorem from Gödel's dissertation, this problem is equivalent to seeking an algorithm for determining whether a particular conclusion may be derived from certain premises using the Hilbert-Ackermann rules of proof. The *Entscheidungsproblem* was called the "principal problem of mathematical logic," because an algorithm for the *Entscheidungsproblem* could, in principle, be used to answer any mathematical question: it would suffice to employ a formalization in first-order logic of the branch of mathematics relevant to the question under consideration. Alan Turing's attention was drawn to the *Entscheidungsproblem* by Newman's lectures, and he soon saw how to settle the problem negatively. That is, Turing showed that no algorithm exists for solving the *Entscheidungsproblem*. The tools that Turing developed for this purpose have turned out to be absolutely fundamental for computer science.

If a positive solution of the *Entscheidungsproblem* would lead to algorithms for settling all mathematical questions, then it must follow that if there is even one problem that has no algorithmic solution, then the *Entscheidungsproblem* itself must have no algorithmic solution. Now, the intuitive notion of *algorithm* serves perfectly well when what we need to verify is that some proposed procedure does indeed constitute a positive solution to a given problem. However, remaining on this intuitive level, we could not hope to prove that some problem has no algorithmic solution. In order to be certain that no algorithm will work, it would appear to be necessary to somehow survey the class of all possible algorithms. This is the task that Turing set himself.

Turing began with a human being who would carry out the successive steps called for by some algorithm; that is, Turing proposed to consider the behavior of a "computer." Here the word *computer* refers to a person carrying out a computation; this was how Turing (and everyone else) used the word in 1935. Turing then proceeded [1936], by a sequence of simplifications each of which could be seen to make no essential difference, to obtain his characterization of computability.

Turing's first simplification was to assume that "the computation is carried out on one-dimensional paper, i.e., on a tape divided into squares" since "it will be agreed that the two-dimensional character of paper is no essential of computation." Turing continued,

> I shall also assume that the number of symbols...used...is finite. ...this restriction...is not very serious. It is always possible to use sequences of symbols in the place of single symbols. ...The behavior of the computer at any moment is determined by the symbols which he is observing, and his 'state of mind' at that moment. We may suppose that there is a bound B to the number of...squares which the computer can observe at one moment. We will also suppose that the number of states of mind which need be taken into account is finite.[2]

Turing next argued that the entire computation can be thought of as a sequence of atomic "simple" steps, each of which consists of:

1. a change of one symbol in one of the B "observed" squares (changes of more than one symbol can always be reduced to successive changes of a single symbol);

2. changes from the squares currently being "observed" to other squares no more than a distance of L squares away, where L is some constant; and

3. a change in the "state of mind" of the "computer."

The final step in Turing's analysis is the crucial remark that the outcome will in no way be altered if the human computer is replaced by a machine capable of a finite number of distinct states, or "*m*-configurations" as Turing called them, corresponding to the different states of mind the human computer possesses in the course of the computation. Each *m*-configuration is then completely characterized by a table which specifies the changes of types

1, 2, and 3 above, corresponding to each possible set of observed squares and symbols appearing in those squares. Hence, any algorithmic process can be carried out by machine, and, moreover, by a machine that conceptually is quite simple. As an addendum to his analysis, Turing pointed out that his machines could be simplified even further, permitting only one square at a time to be observed or "scanned" and allowing only a change of the observed square to the square immediately to the left or immediately to the right. Unlike the previous simplifications, this one can be justified rigorously: one can *prove* that nothing is lost by the restriction to the simpler machines.

We have arrived at the famous notion of a *Turing machine*: a Turing machine can be specified by a finite set of quintuples, each having one of the three forms:

$$p\alpha\beta Rq \quad \text{or} \quad p\alpha\beta Lq \quad \text{or} \quad p\alpha\beta Nq.$$

Such a quintuple signifies that if the machine is in m-configuration p and the symbol α appears in the scanned square, then the machine will replace α by β, enter m-configuration q, and move one square to the right, move one square to the left, or not move at all, depending on whether the third symbol in the quintuplet is R, L, or N, respectively. The quintuples which constitute a particular Turing machine determine the course of a computation. Turing's machines were to be *deterministic* in the sense that no two of its quintuples are allowed to begin with the same pair p, α, so that, at any particular time, no more than one action could be called for. If a machine were to be in m-configuration p scanning a symbol α, such that *no* quintuple of the machine begins with the pair $p\alpha$, then the computation would be at an end—the machine would "halt." At any stage of a computation by a Turing machine, there will be only a finite numer of nonblank squares on the tape. But it is crucial that there be no *a priori* bound on this number. Thus, the tape is infinite, in the sense that additional (possibly blank) squares are always available to the right of the scanned square.

If Turing's analysis is accepted, then it may be concluded that any algorithmic process whatever can be carried out by one of these Turing machines. In particular, if the *Entscheidungsproblem* has an algorithmic solution, then it can be solved by a Turing machine. Moreover, for the very reason that made it appropriate to call the *Entscheidungsproblem*, "the principal problem of mathematical logic," if any problem at all can be shown to be unsolvable by Turing machines, the unsolvability of the *Entscheidungsproblem* should readily follow.

In obtaining his undecidable arithmetic propositions, Gödel had made use of a diagonal method that was formally similar to Cantor's proof that the set of real numbers is not countable. Therefore, it was quite natural that Turing should think of applying diagonalization. He called a real number[3] in the interval (0,1) *computable* if there exists a Turing machine that, beginning with a blank tape, successively prints the infinite sequence of 0s and 1s constituting the number's binary expansion. (Symbols other than 0 and 1 printed by the machine are just ignored for this purpose.) A machine that computes a real number in this sense was called *circle-free*; one that does not (because it never prints more than a finite number of 0s and 1s) was called *circular*. Since there are only countably many distinct finite sets of quintuples, the set of computable real numbers is likewise countable. Yet superficially, it appears that one could argue the reverse. As Turing expressed it, speaking in terms of the *sequences* of 0s and 1s corresponding to computable real numbers:

> If the computable sequences are enumerable, let α_n be the *n*-th computable sequence, and let $\varphi_n(m)$ be the *m*-th figure in α_n. Let β be the sequence with $1 - \varphi_n(n)$ as its *n*-th figure. Since β is computable, there exists a number K such that $1 - \varphi_n(n) = \varphi_K(n)$ all *n*. Putting $n = K$, we have $1 = 2\varphi_K(K)$, i.e., 1 is even. This is impossible. The computable sequences are therefore not enumerable. [Turing, 1936, 246; (reprinted in Davis, 1965, 132)]

Turing continued (the words in brackets are substituted for Turing's for expository reasons):

> The fallacy in this argument lies in the assumption that β is computable. It would be true if we could enumerate the computable sequences by finite means, but the problem of enumerating computable sequences is equivalent to the problem of finding out whether a given [finite set of quintuples determines] a circle-free machine, and we have no general process for doing this in a finite number of steps. In fact, by applying the diagonal process argument correctly, we can show that there cannot be any such general process.

In other words, if there were such a "general process," it could be used to delete non circle-free machines from an enumeration of all possible finite sets of quintuples, thereby producing an enumeration of "the computable sequences by finite means." But then β, as defined above, would be computable, and we would be led to a contradiction. The only way out is to conclude that "there cannot be any such general process."[4] The problem of determining whether the Turing machine defined by a given finite set of quintuples is circle-free has no algorithmic solution! Of course, in this form, the result depends on accepting Turing's analysis of the computation process. However, it is possible to state the result in the form of a rigorously proved theorem about Turing machines. For this purpose, let us imagine the very quintuples constituting a Turing machine themselves placed on a Turing machine tape. We could then seek to construct a Turing machine M which, begun with a set of quintuples defining any particular Turing machine N on its tape, will eventually halt with an affirmative or a negative message on its tape, according as N is or is not circle-free. Turing's argument can then be used to prove that there cannot be such a Turing machine M. (To make this entirely precise, it is necessary to be explicit as to how the quintuples, as well as the affirmative and negative output messages, are to be coded on the tape in terms of some finite alphabet. But this causes no difficulty.)

Turing next showed that there is no algorithm for the:

BLANK TAPE PRINTING PROBLEM.   *To determine whether a given Turing machine, starting with a blank tape, will ever print some particular symbol, say |.*

Turing's proof of this result proceeds by showing that if there were such an algorithm, then there must also be an algorithm for determining whether a given Turing machine is circle-free. This argument is a bit complicated,[5] and we outline a simpler proof that uses another diagonalization. First we show that there is no algorithm for the:

GENERAL PRINTING PROBLEM.   *To determine whether a given Turing machine, starting with a given string of symbols on its tape, will ever print |.*

Suppose there were an algorithm for this problem. Then, in particular, there would be an algorithm to determine whether a given Turing machine, *starting with its own set of quintuples on its tape*, will ever print |. So, it would follow from Turing's analysis that a Turing machine M could be constructed that would respond to a set of quintuples on its tape by printing | if and only if the machine defined by that set of quintuples *never* prints | when started with its own set of quintuples on its tape. Now, what happens when M is started with *its own set of quintuples* on its tape? It eventually prints | if and only if it never prints |! This contradiction shows that there can be no algorithm for the general printing problem. Finally, if there were an algorithm for the blank tape printing problem, it could also be used to solve the general printing problem. Namely, given the quintuples constituting a Turing machine M together with the string of symbols σ on its tape, we can construct a machine N that, beginning with a blank tape, first prints the string σ, and then behaves exactly like M. So N will eventually print | beginning with a blank tape if and only if M will eventually print | beginning with the string σ on its tape.

Turing used the fact that there is no algorithm for the blank tape printing problem to show that Hilbert's *Entscheidungsproblem* is likewise unsolvable. Turing proceeded as follows: With each Turing machine M, he associated a formula α(M) of first-order logic which, roughly speaking, describes the behavior of M starting with a blank tape. He constructed a second formula β which has the interpretation that the symbol | eventually appears on the tape. It was then not difficult to see that β follows from α(M) by the Hilbert-Ackermann rules if and only if the Turing machine M eventually prints |. Thus, an algorithm for the *Entscheidungsproblem* would lead to an algorithm for the blank tape printing problem.

The notion of Turing machine was developed in order to solve Hilbert's *Entscheidungsproblem*. But it also enabled Turing to realize that it was possible to conceive of a single machine that was capable of performing all possible computations. As Turing expressed it: "It is possible to invent a single machine which can be used to compute any computable sequence." Turing called such a machine *universal*. A Turing machine U was to be called universal if, when started with a (suitably coded) finite set of quintuples defining a Turing machine M on its tape, U would proceed to compute the very same sequence of 0s and 1s that M would compute (beginning with an empty tape). Now, intuitively speaking, there clearly exists an algorithm that does what is required of the universal machine U; the algorithm just amounts to carrying out the instructions expressed by M's quintuples. Thus, the existence of a universal Turing machine is a consequence of Turing's analysis of the concept of computation. On the other hand, it is a rather implausible consequence. Why should we expect a single mechanism to be able to carry out algorithms for "the numerical solution of differential equations" as well as those needed to "make bills for a department store?" However, Turing did not simply depend on the validity of his analysis. He proceeded to produce in detail the actual quintuples needed to define a universal machine. Thus, in the light of the apparent implausibility of the existence of such a machine, Turing was entitled to regard his success in constructing one as a significant vindication of his analysis. The universal machine U actually given by Turing can be thought of as being specified by what is nowadays called an *interpretative* program. U operates by scanning the coded instructions (that is, quintuples) on its tape and then proceeding to carry them out. Of course, many interpretative programs have been constructed in recent years to make it possible to run programs written in such languages as BASIC, LISP, SNOBOL, and PROLOG, but Turing's was the first.

Turing's analysis provided a new and profound insight into the ancient craft of computing. The notion of computation was seen as embracing far more than arithmetic and algebraic calculations. And at the same time, there emerged the vision of universal machines that "in principle" could compute everything that is computable. Turing's examples of specific machines were already instances of the art of programming; the universal machine in particular was the first example of an interpretative program. The universal machine also provided a model of a "stored program" computer in which the coded quintuples on the tape play the role of a stored program, and in which the machine makes no fundamental distinction between "program" and "data." Finally, the universal machine showed how "hardware" in the form of a set of quintuples thought of as a description of the functioning of a mechanism, can be replaced by equivalent "software" in the form of those same quintuples in coded form "stored" on the tape of a universal machine.

While working out his proof that there is no algorithmic solution to the *Entscheidungsproblem*, Turing did not suspect that similar conclusions were being reached on the other side of the Atlantic. In fact, Newman had already received the first draft of Turing's paper when an issue of the *American Journal of Mathematics* arrived in Cambridge containing an article by Alonzo Church of Princeton University, entitled "An Unsolvable Problem of Elementary Number Theory." In this paper, Church had already shown that there were algorithmically unsolvable problems. His paper did not mention machines but it did point to two concepts, each of which had been proposed as explications of the intuitive notion of computability or, as Church put it, "effective calculability." The two concepts were λ-*definability*, developed by Church and his student Stephen Kleene, and *general recursiveness*, proposed by Gödel (in lectures at the Institute for Advanced Study in Princeton during the spring of 1934) as a modification of an idea of J. Herbrand. The two notions had been proved to be equivalent, and Church's unsolvable problem was in fact unsolvable with respect to either equivalent notion. Although in this paper, Church had not drawn the conclusion that Hilbert's *Entscheidungsproblem* was itself unsolvable with respect to these notions, volume 1 (1936), number 1 of the new quarterly *Journal of Symbolic Logic* contained a brief note by Church in which he did exactly that. A later issue of volume 1 of the same journal contained an article by Emil Post, taking cognizance of Church's work, but proposing a formulation of computability very much like Turing's. Turing quickly showed that his notion of computability was equivalent to λ-definability, and he decided to attempt to spend some time in Princeton.

Thus, much of what Turing had accomplished amounted to a rediscovery of what had already been done in the United States. But his analysis of the notion of computation and his discovery of the universal computing machine were entirely novel, going beyond anything that had been done in Princeton. In particular, although Gödel had remained unconvinced by the evidence available in

Princeton, that Church's proposal to identify effective calculability with the two equivalent proposed notions was correct, Turing's analysis finally convinced him.[6]

Turing was at Princeton for two academic years beginning in the summer of 1936. Formally, he was a graduate student, and indeed he did complete the requirements for a doctorate with Alonzo Church as his thesis adviser. His doctoral dissertation was his deep and important paper [Turing, 1939], in which he studied the effect on Gödel undecidability of transfinite sequences of formal systems of increasing strength. This paper also introduced the key notion of an *oracle*, which made it possible to classify unsolvable problems,[7] and which is playing a very important role in current research in theoretical computer science.[8] Some writers have been confused about the circumstances under which Turing was a graduate student at Princeton, and have assumed that Turing's earlier work on computability had been done at Princeton under Church's supervision. A circumstance that may have helped lead to this confusion is that the published account [Turing, 1936] of the work on computability concludes with an appendix (in which a proof is outlined of the equivalence of Turing's concept of computability with Church's λ-definability) dated August 28, 1936 at "The Graduate College, Princeton University, New Jersey, U. S. A."[9]

Fine Hall[10] in 1936 housed not only the mathematics faculty of Princeton University, but also the mathematicians who were part of the recently established Institute for Advanced Study. The great influx to the United States of scientists fleeing the Nazi regime had begun. The concentration of mathematical talent at Princeton during the 1930s came to rival and then surpass that at Göttingen, where David Hilbert held sway. Among those to be seen in the corridors of Fine Hall were Solomon Lefschetz, Hermann Weyl, Albert Einstein, and...John von Neumann.[11] During Turing's second year at Princeton, he held the prestigious Procter Fellowship. Among the letters of recommendation written in support of his application was the following:

June 1, 1937

Sir,

Mr. A. M. Turing has informed me that he is applying for a Proctor [*sic*] Visiting Fellowship to Princeton University from Cambridge for the academic year 1937–1938. I should like to support his application and to inform you

THE ORIGIN OF MODERN COMPUTERS 151

that I know Mr. Turing very well from previous years: during the last term of
1935, when I was a visiting professor in Cambridge, and during 1936–1937,
which year Mr. Turing has spent in Princeton, I had opportunity to observe
his scientific work. *He has done good work in branches of mathematics in
which I am interested, namely: theory of almost periodic functions, and theory
of continuous groups.* [emphasis added]

I think that he is a most deserving candidate for the Proctor [sic]
Fellowship, and I should be very glad if you should find it possible to award
one to him.

I am, Respectfully, John von Neumann   [Hodges, 1983, 131]

Thus, as late as June 1937, either von Neumann was unaware of
Turing's work on computability, or he did not think it appropriate
to mention it in a letter of recommendation. There have been
tantalizing rumors of important discussions between the two
mathematicians about computing machinery, during the Princeton
years, or later, during the Second World War. But there does not
appear to be any real evidence that such discussions ever took
place.[12] However, von Neumann's friend and collaborator Stanis-
law Ulam, in a letter to Andrew Hodges [Hodges, 1983, 145],
mentioned a game that von Neumann had proposed during the
summer of 1938 when he and Ulam were travelling together in
Europe; the game involved "writing down on a piece of paper as
big a number as we could, defining it by a method which indeed
has something to do with some schemata of Turing's."[13] Ulam's
letter also stated that "von Neumann mentioned to me Turing's
name several times in 1939 in conversations, concerning mechanical ways to develop formal mathematical systems." On the basis of
Ulam's letter, it seems safe to conclude that by the outbreak of the
Second World War in September 1939, von Neumann was well
aware of Turing's work on computability and regarded it as important.

When did Turing begin to think about the possibility of constructing a physical device that would be, in some appropriate
sense, an embodiment of his universal machine? According to
Turing's teacher, M. H. A. Newman, this was in Turing's mind
from the very first. In an obituary article in *The Times*, Newman
wrote (quoted in [Hodges, 1983, 545]):

The description that he then gave of a "universal" computing machine was
entirely theoretical in purpose, but Turing's strong interest in all kinds of

practical experiment made him even then interested in the possibility of actually constructing a machine on these lines.

In Princeton, Turing's "practical" interests included a developing concern with cryptanalysis. Possibly in this connection, he designed an electro-mechanical binary multiplier, and gaining access to the Physics Department graduate student machine shop,[14] he constructed various parts of the device, building the necessary relays himself. Another of Turing's interests during this period—an interest which combined the theoretical with practical computation —was the famous Riemann Hypothesis concerning the distribution of the zeros of the Riemann ζ-function. Shortly after Turing returned to England in the summer of 1938, he applied for and was granted £40 to build a special purpose analogue computer for computing Riemann's ζ-function,[15] which Turing hoped to use to test the Riemann Hypothesis numerically [Hodges, 1983, 138–140, 155–158]. But even as Turing was beginning serious work on this machine, the Second World War intervened and moved him in quite another direction. The ζ-function machine was never completed.

TO BUILD A BRAIN

Turing spent the war years at Bletchley Park, a country mansion that housed Britain's brilliant group of cryptanalysts. The Germans had developed improved versions of a commercial encrypting machine, the *Enigma*. The task of breaking the Enigma code fell to the group at Bletchley Park. They were given a head start in their task by having access to the work of a group of Polish mathematicians who had succeeded with an earlier and considerably simpler version of the Enigma. Building on this work, Turing and Gordon Welchman (an algebraic geometer from Cambridge) progressed to the point where machines, called Bombes (the name first used by the Poles for their much more primitive device), could be built to decode everyday German military communications. Naval communications were Turing's special province, and by the summer of 1941 the information derived from the Bombes enabled the British Admiralty to defeat the German submarine offensive against

Atlantic shipping that had been threatening to strangle a be-leagured Britain ([Welchman, 1982] [Hodges, 1983, 160–210]).[16] But this great success was a precarious one. It was clear that if the Germans introduced more complexity into their procedures, the Bombes would be overwhelmed. And so, more ambitious machines (which indeed turned out to be necessary) were constructed: first the Heath Robinson series, and later the Collossus. The latter, constructed in 1943 under the direct supervision of M. H. A. Newman, used vacuum tube circuits to carry out complex Boolean computations very rapidly. The Collossus contained 1500 tubes and was built in the face of skepticism on the part of the engineers that so many vacuum tubes could work together, without a failure, long enough to get useful work done.

Thus, when the war ended, Turing had a solid basic knowledge of electronics, and knew that large scale computing machines could be constructed using electronic circuits. The significance for Turing of this practical knowledge cannot be fully grasped without taking into account the new conceptual framework for thinking about computing to which his work on computability had led him. For Turing had been led to conclude that computation was simply carrying out the steps in some "rule of thumb process" (as Turing expressed it in an address to the London Mathematical Society [Turing, 1947, 107]). A "rule of thumb process" is to be understood as one which can be carried out simply by following a list of unambiguous instructions referring to finite discrete configurations of whatever kind. Turing's work had also shown that, without loss of generality, one could restrict oneself to instructions of an extremely simple kind. Finally, it was possible to construct a single mechanical device capable, in principle, of carrying out any computation whatever. "It can be shown that a single machine...can be made to do the work of all" [Turing, 1947, 112]. As exciting as this prospect must have appeared, it was only part of Turing's remarkable vision. Turing dared to imagine that computation encompassed far more than mere calculation, that it actually included the human mental processes that we call "thought." He was interested in much more than a machine capable of very rapid computation; Alan Turing wanted to build a brain. This vision was the subject of much discussion at Bletchley Park, where Turing focused on chess as an example of human "thought" that should be capable of mechanization. The full scope of Turing's thought was only exposed to the public later, in [Turing, 1947] and in his now classical essay [Turing, 1950]. In 1947 Turing was already speaking of circumstances in which "one is obliged to regard the machine as showing intelligence. As soon as one can provide a reasonably large memory capacity it should be possible to begin to experiment on these lines" [Turing, 1947, 123]. As we shall see, Turing was eager to help build such machines. But for a second time, Turing's professional life was profoundly affected by developments in the Western Hemisphere.

VON NEUMANN AND THE MOORE SCHOOL

As has already been noted, by the summer of 1938 von Neumann was very much aware of Turing's work on computability. There is also evidence that, early on, von Neumann perceived that Turing's work had implications for the practice of computation. A wartime colleague of von Neumann recalled that "in about 1943 or 44 von Neumann was well aware of the fundamental importance of Turing's paper of 1936...and at his urging I studied it...he emphasized...that the fundamental conception is owing to Turing" [Hodges, 1983, 145, 304]. Herman Goldstine (who was von Neumann's close collaborator) said, "There is no doubt that von Neumann was thoroughly aware of Turing's work" [Goldstine, 1972, 174].

As with Turing, von Neumann's wartime work involved large-scale computation. But, where the cryptanalytic work at Bletchley Park emphasized the discrete combinatorial side of computation, so in tune with Turing's earlier work, it was old-fashioned, heavy, number-crunching that von Neumann needed. Although he had tried to inform himself about new developments in computational equipment, von Neumann learned of the ENIAC project quite fortuitously on meeting the young mathematician Herman Goldstine at a railway station during the summer of 1944. Von Neumann quickly became a participant in discussions with the ENIAC group at the Moore School in Philadelphia.

The Collossus with its 1500 vacuum tubes was already an engineering marvel; the ENIAC with 18000 tubes was simply astonishing. The conventional wisdom of the time was that no such assemblage could do reliable work; it was held that the mean free path between vacuum tube failures would be a matter of seconds. It was the chief engineer on the ENIAC project, John Prosper Eckert, Jr., who was largely responsible for the project's success. Eckert insisted on extremely high standards of component reliability. Tubes were operated at extremely conservative power levels, and the failure rate was kept to 3 tubes per week. The ENIAC was an enormous machine, occupying a large room. It was a decimal machine and was programmed by connecting cables to a plugboard [Burks and Burks, 1981], rather like an old-fashioned telephone switchboard.

By the time that von Neumann began meeting with the Moore School group, it was clear that there were no important obstacles to the successful completion of the ENIAC, and attention was focused on the next computer to be built, tentatively called the EDVAC. Von Neumann immediately involved himself with the problems of the *logical* organization of the new machine. As Goldstine [1972, 186] recalls, "Eckert was delighted that von Neumann was so keenly interested in the logical problems surrounding the new idea, and these meetings were scenes of greatest intellectual activity." Goldstine [1972, 188] comments:

> This work on the logical plan for the new machine was exactly to von Neumann's liking and precisely where his previous work on formal logics [*sic*] came to play a decisive role. Prior to his appearance on the scene, the group at the Moore School concentrated primarily on the *technological* problems, which were very great; after his arrival he took over leadership on the *logical* problems.

A key idea emphasized in the meetings was that any significant advance over the ENIAC would require a substantial capacity for the internal storage of information. This was because communication with the exterior would be at speeds far slower than the internal electronic speeds at which the computer could function and, therefore, constituted a potential bottleneck. Once again, John Eckert played a crucial role. He had previously shown how to modify a device called a delay line (originally developed by the physicist W. B. Schockley who later invented the transistor) so as to be a working component of radar systems. These delay lines (which stored information in the form of a vibrating tube of mercury) were just what was needed.

The communication bottleneck would be evident in the case of any computation involving the manipulation of large quantities of data. But it was even more crucial for the *instructions* that the computer would carry out. Indeed, it would make little sense for a computer to produce the results of a calculation rapidly, only to wait idly for the next instruction. The solution was to store the instructions internally with the data: what has come to be called the "stored program concept."

The computers of the postwar period differed from previous calculating devices in having provision for internal storage of programs as well as data. But they were different in another more fundamental way. They were conceived, designed, and constructed, not as mere automatic calculators, but as *engines of logic*, incorporating the general notion of what it means to be computable and embodying a physical model of Turing's universal machine. Whereas there has been a great deal of discussion concerning the introduction of the "stored program concept," the significance of this other great, but rather subtle, advance has not been fully appreciated. In fact, the tendency has been to use the single term, "stored-program concept," to include all of the innovations introduced with the EDVAC design. This terminological confusion may well be responsible, at least in part, for the fact that there has been so much acrimony about who deserves credit for the revolutionary advances in computing which took place at this time (see, for example, the report [Aspray, 1982]).

The key document in which this new conception of computer first appeared was the draft report [von Neumann, 1945] which quickly became known as the EDVAC Report. This report never advanced beyond the draft stage, and is quite evidently incomplete in a number of ways. Yet it was widely circulated almost at once and was very influential. In fact the conception of computing machine it embodies has come to be known as the "von Neumann architecture." One element of controversy, which will probably never be fully resolved, is the question of how much of the EDVAC report represented von Neumann's personal contribution. Although Eckert and his consultant J. W. Mauchly later denied that von Neumann had contributed very much, shortly after the report appeared they wrote as follows:

> During the latter part of 1944, and continuing to the present time, Dr. John von Neumann...has fortunately been available for consultation. He has contributed to many discussions on the logical controls of the EDVAC, has prepared certain instruction codes, and has tested these proposed systems by writing out the coded instructions for specific problems. Dr. von Neumann has also written a preliminary report in which most of the results of earlier discussions are summarized. ...In his report, the physical structures and devices...are replaced by idealized elements to avoid raising engineering problems which might distract attention from the logical considerations under discussion.　[Goldstine, 1972, 191] [Metropolis and Worlton, 1980, 55]

Goldstine [1972, 191–192] (apparently unaware of Turing's claim to be mentioned in this connection) comments:

> Von Neumann was the first person, as far as I am concerned, who understood explicitly that a computer essentially performed logical functions, ...he also made a precise and detailed study of the functions and mutual interactions of the various parts of a computer. Today this sounds so trite as to be almost unworthy of mention. Yet in 1944 it was a major advance in thinking.

One way in which the EDVAC report betrays its unfinished state is by the large number of spaces clearly intended for references, but not filled in. Almost every page contains the abbreviation "cf." followed by a space. All the more significant is the one reference that von Neumann did supply: the reference, supplied in full, was to the paper [McCulloch and Pitts, 1943] in which a mathematical theory of idealized neurons had been developed. Von Neumann suggested that basic vacuum tube circuits could be thought of as physical embodiments of these neurons. Here there are two connections with Turing's ideas. The first, more obvious one, is that like Turing, von Neumann was thinking of a computer as being like a brain (or at least a nervous system). In Ulam's letter to Hodges quoted above, Ulam alluded to this confluence, writing in a postscript: "Another coincidence of ideas: both Turing and von Neumann wrote of 'Organisms' beyond mere computing machines."[17] But a more explicit connection with Turing's work becomes evident on further study. McCulloch himself (see [von Neumann, 1963,

319]) later stated that the paper which von Neumann did reference had been directly inspired by [Turing, 1936]. In fact, the paper itself cites the fact that a universal Turing machine can be modeled in a suitable version of the neural net formalism as the principal reason for believing in the adequacy of the formalism.

There is other evidence that von Neumann was concerned with universality in Turing's sense. Thus he spoke [Randell, 1982, 384] of the "logical control" of a computer as being crucial for its being "as nearly as possible *all purpose*." In order to test the general applicability of the EDVAC, von Neumann wrote his first serious program, not for numerical computation of the kind for which the machine's order code was mainly developed, but rather to carry out a computational task of a logical-combinatorial nature, namely the efficient sorting of data.[18] The success of this program helped to convince von Neumann that "it is legitimate to conclude already on the basis of the now available evidence, that the EDVAC is very nearly an 'all-purpose' machine, and that the present principles for the logical controls are sound" [Goldstine, 1972, 209]. Articles written within a year of the EDVAC report confirm von Neumann's awareness of the basis in logic for the principles underlying the design of electronic computers. The introduction to one such article states:

> In this article we attempt to discuss [large scale computing] machines from the viewpoint not only of the mathematician but also of the engineer and the logician, i.e., of the...person or group of persons really fitted to plan scientific tools. [Goldstine and von Neumann, 1946]

Another article [Burks, Goldstine, and von Neumann, 1946] clearly alludes to Turing's work, even as it indicates that purely logical considerations are not enough:

> It is easy to see by formal-logical methods that there exist codes that are *in abstracto* adequate to control and cause the execution of any sequence of operations which are individually available in the machine and which are, in their entirety, conceivable by the problem planner. The really decisive considerations from the present point of view, in selecting a code, are of a more practical nature: simplicity of the equipment demanded by the code, and the clarity of its application to the actually important problems together with the speed of its handling those problems. It would take us much too far afield to discuss these questions at all generally or from first principles.

There has been much acrimony over the question of just what von Neumann had contributed; indeed, this question even became the subject of extensive litigation. Much of the controversy concerns the relative significance of the contributions of von Neumann on the one hand, and of Eckert and Mauchly on the other. In particular, some recent studies challenge the belief that von Neumann's technical contributions were of much importance. (See the semi-popular history [Shurkin, 1984] and the meticulously researched [Stern, 1981].) It is not difficult to understand why this should be. The Turing-von Neumann view of computers is conceptually so simple and has become so much a part of our intellectual climate, that it is difficult to understand how radically new it was. It is far easier to appreciate the importance of a new invention, like the mercury delay line, than of a new and abstract idea.

THE ACE

Meanwhile, what of Turing? His mother (quoted in [Hodges, 1983, 294]) reports his saying "round about 1944" that he had plans "for the construction of a universal computer." During the war, he had been telling colleagues that he wanted to build a "brain." He proposed to construct an electronic device that would be a physical realization of his universal machine. Early in 1945, while on a trip to the United States, J. R. Womersley, Superintendent of the Mathematics Division of the National Physical Laboratory of Great Britain, was introduced to the ENIAC and to the EDVAC report. As early as 1938, Womersley had considered the possibility of constructing a "Turing machine using automatic telephone equipment." His reaction to what he had learned in the United States was to hire Alan Turing [Hodges, 1983, 306–307]. By the end of 1945, Turing had produced his remarkable ACE report [Turing, 1945]. The excellent article by Carpenter and Doran [1977] contains an analysis of the ACE report, comparing it in some detail with von Neumann's EDVAC report. They note that, whereas the EDVAC report "is a draft and is unfinished...more important...is incomplete" the ACE report "is a complete description of a computer, right down to the logical circuit diagrams" and even including "a cost estimate of £11,200." Not surprisingly, Turing showed that he understood the scope of universality. He suggested that his ACE might be able to play chess and to solve jigsaw puzzles. The ACE report contains explicit mention of features such as an instruction address register and truly random access to memory locations, neither of which is dealt with in the EDVAC report, although both are already to be found in [Burks, Goldstine, and von Neumann, 1946]. Although it is known [Goldstine, 1972, 218] that the ACE report quickly made its way across the Atlantic, it seems impossible to determine whether the ACE report influenced the American developments.

It should also be mentioned that the ACE report showed an understanding of numerous issues in computer science well ahead of its time. Of these perhaps the most interesting are microprogramming and the use of a stack for a hierarchy of subroutine calls. We have already mentioned Turing's address, to the London Mathematical Society early in February 1947, in which he unveiled the scope of his vision regarding the ACE and its successors. In this talk he explained that one of the central conclusions of his earlier work on computability was that "the idea of a 'rule of thumb' process and a 'machine process' were synonymous.... Machines such as the ACE may be regarded as practical versions of...the type of machine I was considering.... There is at least a very close analogy...digital computing machines such as the ACE...are in fact practical versions of the universal machine." Turing went on to raise the question of "how far it is in principle possible for a computing machine to simulate human activities." This led him to propose the possibility of a computing machine programmed to learn and permitted to make mistakes. "There are several theorems which say almost exactly that...if a machine is expected to be infallible, it cannot also be intelligent.... But these theorems say nothing about how much intelligence may be displayed if a machine makes no pretence at infallibility."

Turing was much better at communicating in this visionary manner than he was in dealing with the bureaucrats who actually allocate resources, and he had considerable difficulty in getting his ideas put into practice. However, a machine embodying much of the design in the ACE report was eventually constructed, the Pilot ACE, and a successful commercial machine, the DEUCE, followed.

LOGIC AND THE FUTURE

Since 1945, we have witnessed a number of breathtaking technological developments that have completely altered the physical form and the computational power of computers. I sit writing this essay using a text-editing program running on my personal microcomputer with a memory capacity of over 5 million bits. The "johniac" computers on which I learned to program in the 1950s had a total memory of 41,000 bits and were only affordable by institutions. But the connection between logic and computing continues to be a vital one, and the lesson of universality, of the possibility of replacing the construction of diverse pieces of hardware by the programming of a single all-purpose device continues to be relevant. In fact the very existence of personal microcomputers is the result of this lesson being learned anew in the case of integrated circuit technology. In 1971, faced with the requirement for ever more complex and diverse "chips" by his employer, the Intel Corporation, Marcian Hoff found the solution: a single all-purpose programmable chip, and the microprocessor was born. We can foresee that this will happen again and again as technology continues its march towards faster and smaller components.

NOTES

[1] In his authoritative biography, Andrew Hodges [1983, 132] quotes Turing as having, on at least one occasion, attributed his homosexuality to his childhood in boarding schools in England far from his parents in India. (Hodges himself makes it clear that he does not accept this explanation.)

[2] For a discussion of the significance of this "finite mental states" hypothesis, see [Wang, 1974, 93] and [Webb, 1980, 8 and 221–222].

[3] To include numbers outside the interval (0,1), Turing simply declared any number of the form $n + x$ computable if n is an integer and x is a computable number in the interval (0,1).

[4] Turing recognized that although this proof is "perfectly sound," it "may leave the reader with a feeling that 'there must be something wrong'," and he therefore supplied another proof that does not so closely approach paradox.

[5] A remark for the knowledgeable: the problem of determining whether a given Turing machine is circle-free is complete of degree $0''$, and therefore cannot be reduced to either of the printing problems.

[6] For a discussion of some of the historical issues involved in these developments, as well as references see [Davis, 1982].

[7] Thus suppose that we could, somehow, come to possess an oracle or "black box" which can tell us for a given set of quintuples, whether the Turing machine defined by that set of quintuples is circle-free. Then it is not difficult to show that it is possible to construct a Turing machine which can solve either of the two unsolvable printing problems if only the machine is permitted to ask the oracle questions and make use of the answers. However, this will not work in the other direction. This is expressed by saying that the circle-free problem is of a *higher degree of unsolvability* than the printing problems.

[8] Many important open questions in computer science ask whether certain inclusions between classes of sets of strings are proper. (The famous $P = NP$ problem is of this character.) In many cases (including the $P = NP$ problem), although the original question remains unresolved, it has been proved possible to obtain answers when the problem is modified to permit access to suitable oracles.

[9] Actually even this appendix must have been completed before Turing left England. Turing's departure was on September 23.

[10] Fine Hall in 1936 (and indeed through the 1950s) was a low level attractive red brick building. The building where Princeton's mathematics department is housed today is also called Fine Hall; it is visible as a concrete tower from Highway US 1, a mile away.

[11] Kurt Gödel, who had lectured at the Institute for Advanced Study during the spring of 1934, was unfortunately not in Princeton during Turing's stay. Gödel left Princeton in the fall of 1935 and did not return until after the Second World War had begun.

[12] In the doctoral dissertation [Aspray, 1980, 147–148], there is a reference to discussions between Turing and von Neumann at this time, on the question of whether "computing machines could be built which would adequately model any mental feature of the human brain." Aspray based his account on an interview with J. B. Rosser. However, in a conversation with the present author, Aspray explained that Rosser had not claimed to have himself overheard such discussions, and that Rosser had been unable to remember his source. Aspray indicated that he no longer believes that such a conversation actually occurred. In a recent letter, Alonzo Church indicates that he neither recalled nor could find any record of such "consultations." See also [Randell, 1972].

[13] This sounds very much as though von Neumann had anticipated the important Chaitin-Kolmogoroff notion of descriptive complexity.

[14] The Palmer Physics laboratory was located next door to Fine Hall—there was even a convenient passageway joining the two buildings.

[15] The design of this computer was based on that of a machine in Liverpool which was used to predict the tides.

[16]Another of Turing's contributions to this effort was the invention of new statistical methods for dealing with the vast quantities of data contained in the Enigma "traffic." These methods were later rediscovered independently by the American statistician A. Wald who gave them the name: *sequential analysis*. Surely there cannot be many Britons whose contribution to the ultimate victory approached Turing's. A little over a decade after the turning of the tide in the "Battle of the Atlantic," being duly convicted of performing acts "of gross indecency," Turing was sentenced in a British court of law to a one year probation term, during which time he was required to submit to a course of hormone treatments that amounted to a temporary chemical castration. Such was the hero's reward!

[17]I am grateful to Andrew Hodges for making a copy of this letter available to me.

[18]The sorting algorithm that von Neumann implemented belongs to the family of so-called "merge" sorts. For a very interesting discussion of this program and of the proposed EDVAC order code, see [Knuth, 1970].

Acknowledgement. The research for this article was done while I was a John Simon Guggenheim Fellow (1983–84). The article is part of a comprehensive investigation of the relation between mathematical logic and computation. I am very grateful to the Guggenheim Foundation for its support. I also wish to thank Professors Harold Edwards and Esther Phillips who read an earlier draft and made many helpful suggestions.

REFERENCES

Aspray, William F., 1980, *From Mathematical Constructivity to Computer Science: Alan Turing, John von Neumann and the Origins of Computer Science in Mathematical Logic*, doctoral dissertation, University of Wisconsin.

——, 1982, "History of the Stored-Program Concept," (Report on a session held on this subject as part of Pioneer Day at the National Computer Conference, June 1982), *Annals of the History of Computing*, 4, 358–361.

Burks, A. W., H. H. Goldstine, and John von Neumann, 1946, *Preliminary Discussion of the Logical Design of an Electronic Computing Instrument*, Institute for Advanced Study. Reprinted in [von Neumann, 1963, 34–79].

Burks, Arthur W. and Alice R. Burks, 1981, "The ENIAC: First General-Purpose Electronic Computer," *Annals of the History of Computing*, 3, 310–399.

Carpenter, B. E. and R. W. Doran, 1977, "The Other Turing Machine," *Computer Journal*, 20, 269–279.

——, 1986, *A. M. Turing's Ace Report of 1946 and Other Papers*, M.I.T. Press.

Ceruzzi, Paul E., 1983, *Reckoners, the Prehistory of the Digital Computer, from Relays to the Stored Program Concept, 1933–1945*, Greenwood Press, Westport, Connecticut.

Davis, Martin, 1965, *The Undecidable*, Raven Press.

————, 1982, "Why Gödel Didn't Have Church's Thesis," *Information and Control*, 54, 3–24.

————, 1983, "The Prehistory and Early History of Automated Deduction," in *Automation of Reasoning*, vol. 1, Jörg Siekmann and Graham Wrightson (editors), Springer-Verlag.

Goldstine, Herman H. and John von Neumann, 1946, *On the Principles of Large Scale Computing Machines*, unpublished. Reprinted in [von Neumann, 1963, 1–32].

Goldstine, Herman H., 1972, *The Computer from Pascal to von Neumann*, Princeton University Press.

Hilbert, D. and W. Ackermann, 1928, *Grundzüge der Theoretischen Logik*, Julius Springer.

Hodges, Andrew, 1983, *Alan Turing: The Enigma*, Simon and Schuster.

Huskey, V. R. and H. D. Huskey, 1980, "Lady Lovelace and Charles Babbage," *Annals of the History of Computing*, 2, 299–329.

Kneale, W. and M. Kneale, 1962, *The Development of Logic*, Oxford.

Knuth, D. E., 1970, "Von Neumann's First Computer Program," *Computer Surveys*, 2, 247–260.

McCulloch, W. S. and W. Pitts, 1943, "A Logical Calculus of the Ideas Immanent in Nervous Activity," *Bull. Math. Biophys.*, 5, 115–133. Reprinted in McCulloch, W. S., *Embodiments of Mind*, M.I.T. Press, 1965, pp. 19–39.

Metropolis, N., J. Howlett, and Gian-Carlo Rota (editors), 1980, *A History of Computing in the Twentieth Century*, Academic Press.

Metropolis, N. and J. Worlton, 1980, "A Trilogy of Errors in the History of Computing," *Annals of the History of Computing*, 2, 49–59.

Parkinson, G. H. R., 1966, *Leibniz—Logical Papers*, Oxford.

Post, Emil, 1936, "Finite Combinatory Processes. Formulation I," *Journal of Symbolic Logic*, 1, 103–105. Reprinted in [Davis, 1965].

Randell, Brian, 1972, "On Alan Turing and the Origins of Digital Computers," in *Machine Intelligence*, 7, B. Meltzer and D. Michie (editors), Edinburgh, 3–20.

————, 1977, "Colossus: Godfather of the Computer," *New Scientist*, 73, 1038, 346–348. Reprinted in [Randell, 1982, 349–354].

————, (editor), 1982, *The Origins of Digital Computers*, *Selected Papers* (third edition), Springer-Verlag.

Shurkin, Joel, 1984, *Engines of the Mind: A History of the Computer*, W. W. Norton & Company.

Smith, David Eugene, 1929, *A Source Book in Mathematics*, McGraw-Hill.

Stern, Nancy, 1980, "John von Neumann's Influence on Electronic Digital Computing, 1944–1946," *Annals of the History of Computing*, 2, 349–362.

————, 1981, *From Eniac to Univac: An Appraisal of the Eckert-Mauchly Machines*, Digital Press.

Turing, Alan, 1936, "On Computable Numbers with an Application to the Entscheidungsproblem," *Proc. London Math. Soc.*, ser. 2, 42, 230–267. Correction: ibid, 43(1937), 544–546. Reprinted in [Davis, 1965, 116–154].

————, 1939, "Systems of Logic Based on Ordinals," *Proc. London Math. Soc.*, series 2, 45, 161–228. Reprinted in [Davis, 1965, 155–222].

_____, 1945, *Proposals for Development in the Mathematics Division of an Automatic Computing Engine* (*ACE*), 1945, Report E882 National Physical Laboratory of Great Britain. Reprinted 1972 with a forward by D. W. Davies as National Physical Laboratory Report, Com. Sci. 57. Reprinted in [Carpenter and Doran, 1986], 20–105.

_____, 1947, *Lecture to the London Mathematical Society*, first published in [Carpenter and Doran, 1986].

_____, 1950, "Computing Machinery and Intelligence," *Mind*, vol. LIX, No. 236.

von Neumann, John, 1945, *First Draft of a Report on the EDVAC*, Moore School of Electrical Engineering, University of Pennsylvania, unpublished. Reprinted in [Stern, 1981, 177–246].

_____, 1961, *Collected Works*, vol. 1, A. H. Taub (ed.), Pergamon Press.

_____, 1963, *Collected Works*, vol. 5, A. H. Taub (ed.), Pergamon Press.

Wang, Hao, 1974, *From Mathematics to Philosophy*, Humanities Press.

Webb, Judson C., 1980, *Mechanism, Mentalism, and Metamathematics: An Essay on Finitism*, D. Reidel.

Welchman, Gordon, 1982, *The Hut Six Story*, McGraw-Hill.

Woodger, M., 1958, "The History and Present Use of Digital Computers in the National Physical Laboratory," *Process Control and Automation*, November 1958, 437–442. Reprinted with a correction in [Carpenter and Doran, 1986, 125–140].

THE MATHEMATICAL RECEPTION OF THE MODERN COMPUTER: JOHN VON NEUMANN AND THE INSTITUTE FOR ADVANCED STUDY COMPUTER

William Aspray

1. INTRODUCTION

Today the close connection between mathematics and computing is readily apparent. Joint faculty appointments and unified departments of mathematics and computer science provide institutional confirmation, as does the common use of the computer in research in number theory, finite algebra, and mathematical physics and chemistry. The computer has invigorated and redirected research in combinatorics and numerical analysis, and stimulated the growth of such new mathematical disciplines as computational complexity, automata and formal language theory, and continuous information theory. Even in traditional areas of algebra and analysis, where the computer does not figure prominently, there has been a trend to replace abstract existence proofs by computer-assisted constructions. In the other direction, methods and results from all areas of mathematics, but especially from discrete mathematics, logic, and combinatorics, have been applied to understand and support the foundations of computer algorithms, architecture, and programming languages.

In the decade following World War II, the first years of the modern computer, the close relationship between computing and mathematics was not clearly established. The mathematical community was not universal in its recognition or acceptance of the computer as a mathematical tool, although the number of mathematicians using the computer for their research grew rapidly in this period.

This paper studies an important example of the relationship between mathematics and computing in the period 1945–1955: the construction of a computer at the Institute for Advanced Study (IAS) in Princeton, New Jersey under the direction of John von Neumann, and the reception and use of the computer by the international mathematical community. The design and use of this computer has placed it among the most influential stored-program computers. It was housed in perhaps the most distinguished mathematics research center of its day. Von Neumann was one of the first to appreciate the relationship between computing and mathematics and was able to advance this relationship through his concurrent research in automata theory, computer design, numerical analysis, and mathematical physics, his promotion of scientific applications of the IAS computer, and his advocacy of computers for military research.

2.   COMPUTING AND MATHEMATICS PRIOR TO 1940

Automated calculation dates from the seventeenth century, when Schickard, Leibniz, Pascal, and others invented the first machines to carry out arithmetical operations. This technology evolved slowly and only in the late nineteenth century did it become economically viable to manufacture desk calculators. This industry flourished in the first half of the twentieth century and introduced many innovations such as automatic division, addition of a printer, improvements in drive and carry mechanisms, and electromechanical power. But because of their slow speed, their inability to compute sequences of operations automatically, and their lack of storage for data and intermediate results, desk calculators were used sparingly by scientists and engineers.

The second major calculating technology of the early twentieth century is the punched card system, consisting of punches, sorters, tabulators, and printers. The first punched card equipment was built by Herman Hollerith to organize data from the 1890 U.S. Census. Commercial uses were soon identified, and a commercial market supplied by IBM and Remington Rand emerged in the 1920s and 1930s. The ready availability and increased capability of punched card equipment in the 1930s led Wallace Eckert of Columbia University and L. J. Comrie of the Nautical Almanac Office of Greenwich, England to pioneer its scientific application. They used the equipment for harmonic analysis, integration of differential equations, and calculation of astronomical tables. Punched card equipment had the advantages over desk calculators of being able to store almost unlimited amounts of initial data and intermediate results on cards, and to carry out predetermined sequences of operations. Even so, punched card equipment was not used extensively in scientific computation prior to the second world war.

The third major calculating technology of this period were the analog calculating devices. The most notable example was the differential analyzer, but also important were harmonic analyzers, network analyzers, gunnery computers, and analog models such as electrolytic tanks, wind tunnels, and model ship basins. Each of these devices had a specific and narrow application, most often associated with one of the newly mathematized engineering disciplines. MIT was a leader in the development of analog calculating devices, building several differential analyzers to solve differential equations, the Wilbur machine to solve systems of linear equations, and the product intergraph to carry out numerical integration of products of functions (such as those occurring in Fourier analysis). While analog calculators served specific engineering needs well, their range of application was limited: they were difficult to program for new types of calculations and, because they measured rather than counted, their precision was circumscribed. During the 1950s analog computers continued in engineering and some scientific use, but by the late 1960s their inherent limitations led to their replacement by digital computers.

In recent years mathematics has assisted computer development in two ways: by examining the nature and power of computers (e.g., switching theory, formal language theory) and by advancing the knowledge of algorithms and algorithmic methods for use of computers (e.g., numerical analysis, recursive function theory). However, the organizational complexity of these three prewar calculating technologies was not sufficiently great to require or attract mathematical study. Some algorithmic techniques for machine computation were advanced—especially for punched card equipment. But these were only minor improvements in the classical methods of numerical analysis, like Gaussian elimination for systems of linear equations, techniques for solving ordinary differential equations, and Fourier analysis. The most notable mathematical study of algorithms before the Second World War was the work in the 1930s on recursive functions and the nature of effectively computable functions carried out by Gödel, Church, Kleene, Post, and Turing. Their work did not originate in problems of practical computation, but instead evolved from questions of logic and foundations of mathematics. Only after (or possibly during) the war was the relation between recursive function theory and practical computing explored.

3. WORLD WAR II AND COMPUTING

World War II was an important stimulus to the development of both analog and digital computers in England and the United States. Analog computers built during the war, e.g., the Rockefeller Differential Analyzer at MIT and the Bell Telephone Laboratories M-9 antiaircraft gunnery computer, were devised for special purposes and have little relevance to this story. Digital calculating equipment built during the war falls into two categories. One type was used for numerical processing that could not be handled by desk calculators or punched card equipment. The most important example is the ENIAC, built at the University of Pennsylvania under contract for U.S. Army Ordnance (Aberdeen, Maryland) and used for preparing ballistics and other mathematical tables, and for approximate numerical solutions of differential equations in aircraft and atomic energy applications. The other type was used for logical and symbol processing. The prime example is the Colossus built by the British Government Code and Cypher School at Bletchley Park for use in breaking German codes.

These two traditions in digital computing continued after the war, as exemplified by the work of the computer pioneers Alan Turing and John von Neumann. Turing, who gained his practical computing experience at Bletchley Park, was mainly interested in building and using computers for symbolic and combinatoric processing tasks such as game playing, language learning, cryptanalysis, theorem proving, and artificial intelligence. Von Neumann, affiliated with ENIAC during the war, wanted computers to solve large numerical problems associated with mathematical physics and numerical meteorology.

4. JOHN VON NEUMANN AND THE INSTITUTE FOR ADVANCED STUDY

In 1933 John von Neumann, a young but established mathematician with fundamental accomplishments in mathematical logic and the mathematical foundations of quantum mechanics, was appointed as one of the original professors of the Institute for Advanced Study School of Mathematics. The mission of the Institute was to bring together in an environment free of teaching and routine faculty duties, distinguished research mathematicians and promising young Ph.D.s in order that they might devote all their attention to the advancement of the frontiers of mathematics. Together, the Institute's School of Mathematics and the Princeton University mathematics department, which shared quarters on campus throughout the 1930s, represented one of the most distinguished mathematics research centers in the world. Mathematicians in permanent residence in Princeton included James Alexander, Salomon Bochner, Alonzo Church, Albert Einstein, Luther Eisenhart, Kurt Gödel, Solomon Lefschetz, Marston Morse, Oswald Veblen, and Herman Weyl.

About the time that von Neumann joined the Institute for Advanced Study he turned his attention to fluid dynamics. In 1937 von Neumann became a consultant to the Army Ordnance Department at Aberdeen, Maryland, at the urging of his Institute colleague, Oswald Veblen, who had been an officer at Aberdeen during the first world war and continued as a consultant thereafter. At Aberdeen, von Neumann learned from R. H. Kent the related theories of shock and detonation as they were then known, so that by the time of Pearl Harbor in 1941 he was a leading expert in the subject. This expertise led to von Neumann's involvement with a number of government agencies during the war: as a member of the National Defense Research Committee from 1941; as a consultant to the Navy Bureau of Ordnance from 1942; and as an active participant on the Manhattan Project at Los Alamos Laboratory from 1943 to 1945.

Many of the wartime applications encountered by von Neumann and others were intractable by analytical methods; these included continuum dynamics, classical electrodynamics, hydrodynamics, and the theories of elasticity and plasticity. Von Neumann's research in hydrodynamics, for example, involved solving nonlinear partial differential equations for which there were no known analytical methods of solution, or even insight into the character of the problems.

> Our present analytical methods seem unsuitable for the solution of the important problems arising in connection with nonlinear partial differential equations and, in fact, with virtually all types of nonlinear problems in pure mathematics. The truth of this statement is particularly striking in the field of fluid dynamics. Only the most elementary problems have been solved analytically in this field. Furthermore, it seems that in almost all cases where limited successes were obtained with analytical methods, these were purely fortuitous, and not due to any intrinsic suitability of the method to the milieu.... The advance of analysis is, at this moment, stagnant along the entire front of nonlinear problems. That this phenomenon is not of a transient nature but that we are up against an important conceptual difficulty is clear from the fact that, although the main mathematical difficulties in fluid dynamics have been known since the time of Riemann and of Reynolds, and although as brilliant a mathematical physicist as Rayleigh has spent the major part of his life's effort in combating them, yet no decisive program has been made against them—indeed hardly any progress which could be rated as important by the criteria that are applied in other, more successful (linear!) parts of mathematical physics. [Goldstine and von Neumann, 1946, *John von Neumann Collected Works*, V, 2–3]

During the war, von Neumann's knowledge of fluid dynamics led to his involvement in the atomic bomb project at the Los Alamos Scientific Laboratory. In 1943 he joined the group as a consultant to work on the problem of implosion. A major obstacle was the lack of a way to use implosion to make radioactive isotopes of plutonium rapidly attain a critical state. Von Neumann was able to model the implosion problem mathematically, but was stymied in calculating a numerical solution. The battery of "computers," meaning those (usually) women who worked at desk calculators, could not handle the large number of required calculations.

Von Neumann toured computer installations at Harvard University and Bell Laboratories in search of technology to assist his project; but the machines he saw did not meet his needs. In August 1944, quite by accident, he heard about a high speed, electronic calculator, the ENIAC, being built at the Moore School of Electrical Engineering of the University of Pennsylvania. When he first visited the Moore School several weeks later, plans for ENIAC were already complete and construction was underway.

By the time that ENIAC reached working order, near the end of 1945, von Neumann appreciated its potential value to scientific research. He arranged for the first substantial program run on ENIAC to be a simulation by Stanley Frankel and Nicholas Metropolis of the Theoretical Physical Division at Los Alamos. Their program, which required transferring a million IBM cards from the Los Alamos card library to the Moore School, was completed in early 1946 after weeks of computation.

ENIAC received many scientific uses. The Bureau of Ordnance prepared ballistics tables and solved other ordnance problems. With von Neumann's encouragement, Los Alamos physicists used the calculator to test their ideas. The Army offered university scientists free access to ENIAC, prompting a wide range of scientific applications [Goldstine, 1972, 157–166]. During 1946, for instance, ENIAC was used by Hans Rademacher and Harry Huskey for the study of round-off errors, by Frankel and Metropolis for a calculation on the liquid drop model of fusion, by Douglas Hartree for a study of the boundary layers in a compressible fluid flow, by Abraham Taub and Adele Goldstine for an examination of the properties of shock waves, by John Goff to study the thermodynamical properties of gases, by D. H. Lehmer for the investigation of some problems in number theory, and by von Neumann for research in numerical meteorology.

Long before ENIAC ran its first program the Moore School staff had understood that the limited primary storage capacity, the cumbersome method of external coding, and the extensive use of vacuum tubes severly limited the machine's applicability. As the war neared its end and the construction of ENIAC reached a more routine development stage, the Moore School staff undertook negotiations to extend their government contract so as to build an improved version of ENIAC that would alleviate the worst design deficiencies. The government saw merit in this proposal and awarded an extension to build the Electronic Discrete Variable Arithmetic Computer (EDVAC). Plans called for a stored program computer with large mercury delay line storage and a greatly reduced number of vacuum tubes. J. Prosper Eckert and his engineering staff began work immediately on a new storage device, while von Neumann and others planned the logical structure for the machine.

In June 1945 von Neumann prepared a "First Draft of a Report on the EDVAC." It presented the first description of the stored program concept and explained how a stored-program computer processes information. Using terminology from the research of Warren McCulloch and Walter Pitts on neural networks in the human brain (i.e., terms like "organ" and "neuron"), von Neumann was able to describe the stored-program computer in a way that illuminated the logical functions of the necessary constituent parts without having to consider the particular engineering components that might be employed in building them. In this way, he also highlighted the analogy between the stored-program computer and the human brain. The report was widely circulated and studied in the 1940s.

5. PLANNING FOR A COMPUTER AT THE INSTITUTE

In 1945 von Neumann decided that he wanted a computer devoted entirely to the advancement of the mathematical sciences. He made inquiries, hoping to convince the Institute for Advanced Study to support the construction of a computer. However, the absence of laboratories at the Institute and the theoretical bent of its members made the prospects appear unlikely. Mathematics had been chosen as the first discipline for study at the Institute expressly because: it was fundamental; it required no laboratory equipment or support staff; and there was strong consensus in the professional community of what constituted good mathematics and who were the best mathematicians. The permanent faculty had not changed significantly since the Institute's founding, so it is understandable that von Neumann's plan was met with, at best, indifference.

In March 1945 Norbert Wiener wrote von Neumann [Letter, 24 March 1945, *von Neumann Papers*, Library of Congress] to inquire whether he would be interested in accepting a position as head of the mathematics department at the Massachusetts Institute of Technology if computer facilities could be made available to him. Von Neumann used this letter in a campaign to convince the Institute that, if computing facilities were not made available, he would go elsewhere. He strengthened his case by soliciting additional offers of faculty appointment from the University of Chicago and Columbia University. By late 1945, with assistance from Oswald Veblen, von Neumann secured the support of Institute Director Frank Aydelotte and initial planning began. Thereafter, Aydelotte was an unflagging supporter of the computer project.

Originally, the Institute was to be joined in this venture by Princeton University and the Radio Corporation of America (RCA), which had facilities in the Princeton area capable of conducting research and developing storage devices. Von Neumann was concerned that competition from EDVAC would make military funding difficult to secure, so he approached the Rockefeller Foundation. Their support never materialized; but the government's interest in having a computer for scientific research and in testing alternative computer designs was sufficient to garner its support.

Von Neumann's contract with the government called for an experimental scientific instrument. As von Neumann wrote to Admiral Bowen of the Navy Office of Research and Inventions:

> The performance of the computer is to be judged by the contribution which
> it will make in solving problems of new types and in developing new

methods. In other words, it is a scientific tool, to be used in research and experimental work, and not in production jobs. [Letter, 23 January 1946, *von Neumann Papers*, Library of Congress]

Von Neumann hoped that additional machines would be built to carry out computations that the IAS machine had shown to be feasible, thereby preserving the Institute computer for scientific experimentation.

Thus, if a new type of problem arises which cannot be handled on other existing computing machines, and which this computer may seem likely to solve, then the new computer should be used on the problem in question until a method of solution is developed and tested—but it should not be used to solve in a routine manner further problems of the same type. The policy should rather be, to have further electronic computers of this new type built, which will belong, say, to the Navy, and do the routine computing jobs. The Institute's computer should be reserved for the developing and exploring work as outlined on the preceding page [above], and this should be the objective of the envisioned project and the Institute's function. [Letter, 23 January 1946, von Neumann to Bowen, *von Neumann Papers*, Library of Congress]

This plan was followed, and a number of close copies were built with government funds: AVIDAC at Argonne National Laboratory, ILLIAC at the University of Illinois, JOHNNIAC at the Rand Corporation, MANIAC at Los Alamos Scientific Laboratory, ORACLE at Oak Ridge National Laboratory, and ORDVAC at Aberdeen Proving Grounds.

Von Neumann outlined criteria for the selection of problems to program for the Institute computer. They are similar to problems he had hoped to solve on the ENIAC and their selection was governed by scientific interests.

It is clear that the problems to be put on the computer for development and exploration of methods should be judged by fairly general viewpoints, namely: Are they of a new type, not amenable to handling by existing methods and machines? Are they typical of a wide and important evolution of applied mathematics, in the direction of the major interests and needs of the present or the immediately forseeable future? In other words, the criteria have to be scientific, in the meaning of this term in applied mathematics and mathematical physics. [Letter, 23 January 1946, von Neumann to Bowen, *von Neumann Papers*, Library of Congress]

While the advancement of scientific computation was foremost among von Neumann's considerations, he was also aware that, politically, this was probably the only arrangement to which the Institute would agree. Aydelotte was able to convince the Institute Board to approve the computer project on the grounds that it was a scientific project. He likened the computer to the large-scale telescope, whose position as an experimental instrument important to theoretical scientific advancement was widely appreciated.

> I think it is soberly true to say that the existence of such a computer would open up to mathematicians, physicists, and other scholars areas of knowledge in the same remarkable way that the two-hundred-inch telescope promises to bring under observation universes which are at the present moment entirely outside the range of any instrument now existing.   (Minutes of the Regular Meeting of the Board of Trustees, Institute for Advanced Study, 19 October 1945 as quoted in [Goldstine, 1972, 243].)

Aydelotte further emphasized the prestige that would redound from having the premier computing device in the country attract "scientists from all over the country" to conduct research.

> This means, of course, that it would be the most complex research instrument now in existence. It would undoubtedly be studied and used by scientists from all over the country. Scholars have already expressed great interest in the possibilities of such an instrument and its construction would make possible solutions of which man at the present time can only dream. It seems to me very important that the first instrument of this quality should be constructed in an institution devoted to practical applications.   (Minutes of the Meeting of the Committee on the Electronic Computer held in the Director's office, Institute for Advanced Study, November 1945, as quoted in [Goldstine, 1972, 243–244].)

In a letter to M. H. A. Newman, the Manchester mathematician who headed the Colossus and the Manchester Mark I computer projects, von Neumann amplified on the use he envisioned for the Institute computer:

> We propose to use it as a "scientific exploration tool," i.e., in order to find out what to do with such a device. That is, I am convinced, that the methods of "approximation mathematics" will have to be changed very radically in order to use such a device sensibly and effectively—and to get into the position of being able to use still faster ones. I think that the problem in this respect is partly logical and partly analytical, since finding suitable approximation methods and finding and coding the proper machine "setups"

may be the main bottleneck. We want to do a good deal of mathematical and logical work in parallel with the engineering development, a good deal more, with the machine as the "experimental tool," when the machine is ready. [Letter, 19 March 1946, *von Neumann papers*, Library of Congress]

In the end the Institute Board approved the project, and funding was contributed by the Army, Navy, Atomic Energy Commission, and the Institute for Advanced Study. Von Neumann recognized the practical and political difficulties of carrying out extensive hardware development inside the Institute. So, RCA was contracted to develop the electrostatic storage device and the National Bureau of Standards to provide the input and output equipment. There were problems with the storage device, and a modified version first used on the Manchester University Mark I computer was adopted instead. Considering the novelty of the technology, overall construction proceeded smoothly and the computer was placed in operation in 1952.

The IAS computer was a parallel, asynchronous, binary computer consisting of 2300 vacuum tubes. It had an electrostatic Williams tube primary memory of 1024 40-bit words and an auxiliary magnetic drum memory. Punched cards were used for input and output.

6.   USES OF THE COMPUTER

With support from the Office of Naval Research, von Neumann was able to assemble over the period 1945–1955 a group of mathematicians interested in the computer. The group included members of the computer project (Arthur Burks, Herman Goldstine), other members of the Institute (Valentine Bargmann, Deane Montgomery), Princeton University mathematicians (Emil Artin, Salomon Bochner), and many visiting mathematicians (Andrew Booth, Benoit Mandelbrot, Nicholas Metropolis, F. J. Murray, C. V. L. Smith, Abraham Taub, John Todd, and Olga Taussky Todd) [Goldstine, 1972, 292].

Many kinds of problems in applied mathematics and a few problems in pure mathematics were programmed for the Institute computer. Von Neumann was a collaborator on many of these

TABLE 1
Typical Problems Programmed for the Institute
for Advanced Study Computer

Investigator(s)	Problem	Reason(s) for Study
(E. Artin), H. H. Goldstine, and J. von Neumann	Kummer conjecture on ideals	intrinsic value
V. Bargmann, D. Montgomery, and J. von Neumann	method for solving large systems of linear equations	intrinsic value
G. Birkhoff and J. von Neumann	computation of conformal maps and solutions of integral equations	intrinsic value
S. Chandrasekhar and J. von Neumann	gravitational field arising from a random distribution of stars	computation of Bessel and Cylinder functions
H. H. Goldstine and J. von Neumann	matrix inversion	intrinsic value, modern error analysis
E. and D. Lehmer	number theory	intrinsic value
F. J. Murray, H. H. Goldstine, and J. von Neumann	fundamental invariants of conic sections in n-dimensional space subjected to rotations of coordinate axes	method for finding eigenvalues of symmetric matrices
R. D. Richtmyer and J. von Neumann	flow of compressible fluid shock waves	intrinsic value, heuristic methods
M. Schwarzschild	internal structure of stars	numerical integration of systems of differential equations

projects, some of the most important of which are listed in Table 1. These problems were chosen either for their inherent mathematical or scientific importance, or in order to investigate numerical methods.

Meanwhile, von Neumann carried out his own research on the interface between mathematics and computing, which fell into three areas: development of numerical methods suited to the computer, use of the computer as a heuristic tool in pure and applied mathematics, and the mathematical foundations of computing science. These are investigated in the following sections.

7. NUMERICAL METHODS FOR THE COMPUTER

Numerical mathematics advanced rapidly in both the theoretical and practical realms in the nineteenth and early twentieth centuries. Practical methods were developed for solving systems of linear equations, inverting matrices, approximating integrals, and solving ordinary differential equations. Von Neumann saw the computer as presenting an opportunity to make existing numerical methods more effective and to extend numerical techniques to previously inaccessible areas, like partial differential equations describing shock waves in bomb design, supersonic flow in aircraft design, and gas-fluid flow in oil field modeling. In the case of a differential equation, for example, the objective of numerical mathematics is to give an approximate solution to a differential equation that is difficult or impossible to solve exactly. The method is to introduce a system of equations (approximating the exact equation) that can be solved to any accuracy, $\varepsilon > 0$, in a finite number of arithmetic operations, and to find an algorithmic procedure for performing these calculations at practical cost.

As von Neumann recognized, what constitutes practical cost depends on the "inner economy" of the calculating device. When computations are done by hand or with the aid of a desk calculator or punched card equipment, it is considerably more costly—in time expended—to multiply than it is to add or to subtract. Computers have an entirely different inner economy. Multiplications are still more costly than additions, but the computer is so fast (10^5 times faster than any punched card equipment) that multiplication is not prohibitively expensive. However, because of the limited amount of internal storage (the Institute computer had only 1000 words of fast internal storage) and the slow speed at which data is passed from the storage unit to the arithmetic unit (in comparison to the electronic speed of arithmetic operations), it is the storage of initial data and intermediate results that cost relatively more in the computer.

Thus von Neumann believed it was important to begin to adapt or develop new numerical methods that are better suited to the computer; stated simply, ones that limit storage requirements (perhaps) at the expense of additional multiplications. So, he and his colleagues turned their attention to two of the most common and important problems of numerical mathematics: solving large systems of linear equations and inverting large matrices. Large systems of linear equations appear in statistics, electrical networks, and in the intermediate stages of numerical solutions of linear differential and integral equations. Matrix inversion appears equally commonly. The statistician Harold Hotelling called into question the conventional method of Gaussian elimination for solving systems of linear equations by computer. In response, Valentine Bargmann, Deane Montgomery, and von Neumann wrote a paper [1946] modifying the conventional method by adjusting the required number and precision of arithmetical operations so as to improve accuracy and efficiency. Goldstine and von Neumann continued these studies in two papers: in the first [1947a] they presented rigorous error estimates for a method (suitable for the computer) of inverting matrices of high order; and in the second [1951] they used probabilistic methods to replace worst-case error bounds by average case bounds. These were among the first of many results to appear in the 1950s and 1960s in the new, computer-oriented numerical analysis.

Von Neumann was interested not only in modifying existing algorithms to make them work more efficiently with the computer, but also in using the computer to extend the range of numerical mathematics. Successful numerical methods for ordinary differential equations had been introduced in the nineteenth and early twentieth centuries. But new applications of the 1940s and 1950s, like atomic energy, aircraft design, and petroleum exploration, required numerical methods for partial differential equations—for which there had been only limited progress before the second world war (see Birkhoff [forthcoming]). The two major prewar contributions in this area are the 1928 study by Courant, Friedrichs, and Lewy on numerical stability and the relaxation method developed in the 1930s by Southwell. Since von Neumann's contributions followed most closely on the stability work, we restrict our attention to that study.

Courant, Friedrichs, and Lewy showed that mathematicians investigating numerical methods must worry about the vexing problem of numerical stability and that there need not be any absolute relationship between the solution to a partial differential equation and the solution of the difference equations approximating it. They took as an example the equations describing a compressible, non-viscous flow. In their numerical solution the continuous time variable was represented in the conventional way by a succession of discrete time intervals Δt, and the continuous space variable by a succession of discrete space intervals Δx. They demonstrated that the ratio $\Delta t / \Delta x$, known later as the *Courant ratio*, had to be less than or equal to a fixed number c for the approximate solutions to converge to the solution of the exact equation. In the example cited in their paper, this "Courant condition" expressed the physical restraint that a sound wave cannot travel more than Δx distance in Δt seconds.

In lectures given at the Los Alamos Laboratory and the White Oak Naval Ordnance Laboratory in 1947, von Neumann extended the result of Courant, Friedrichs, and Lewy to a wider range of partial differential equations, including some that describe military and commercial applications important in the postwar period. The Courant-Friedrichs-Lewy paper gave stability results for general quasilinear hyperbolic and parabolic differential equations in one independent and two dependent variables. Von Neumann extended their result to higher-order equations with more independent and dependent variables; however, he gained this generality at the sacrifice of rigor. He recognized that his technique was heuristic and argued for its value on empirical grounds, pointing out that "it has not led to false results so far." [von Neumann, 1948a, editorial note]. He applied this technique to great effect in several papers ([von Neumann and Richtmyer, 1947], [Hyman, 1947], and [von Neumann, 1948a; 1948b]) and the technique was used in the solution of the heat flow equation by O'Brien, Hyman, and Kaplan [1950].

8.   THE COMPUTER AS A HEURISTIC MATHEMATICAL TOOL

In addition to adapting numerical methods to the computer, von Neumann was interested in applying the computer to research in pure and applied mathematics. Examples of his application of the computer to the theory of partial differential equations, algebraic number theory, and statistics are given below.

Von Neumann sometimes used the computer in the way calculating equipment had traditionally been used by mathematical scientists, to provide complete computational solutions to mathematical problems. However, he also hoped to employ the computer as a heuristic tool in mathematical research. Used in this way the computer would serve as an experimental instrument, providing guidelines and insights rather than making final calculations or contributing to rigorous proofs.

In the late 1930s and early 1940s, in their investigations in continuum mechanics and mathematical statistics, mathematicians were increasingly confronted with nonlinear partial differential equations for which there was no known method of analytic solution. Von Neumann believed that a new approach to nonlinear problems was needed. He started with the observation that, in approaching the mathematics of the hydrodynamical problems of turbulence, one did not always need to solve even the simplest or most physically obvious cases; instead it was "the common statistical characteristics of the entire family [of problems which] contain the really important insights" [Goldstine and von Neumann, 1946, 3].

Physical insight, rather than a collection of specialized techniques for solving individual cases, was what he believed was required. Von Neumann noted that the mathematical subjects related to elliptic differential equations (e.g., potential theory, conformal mapping, and minimal surfaces) had advanced in just this manner, through the physical insights of Riemann and Plateau.

> Such advances as have been made in the theory of nonlinear partial differential equations, are also covered by this principle, just in what seems to us to be the most decisive instances. Thus, although shock waves were discovered mathematically, their precise formulation and place in theory and their true significance has been appreciated primarily by the modern fluid dynamicists. The phenomenon of turbulence was discovered physically and is still largely unexplored by mathematical techniques. [Goldstine and von Neumann, 1946, 4]

As von Neumann observed, physical experiments had enabled fluid dynamicists to compute approximate answers to very special cases of their problems; this in turn provided insight and offered clues to determining analytic solutions. The instrument of the fluid dynamicist was the wind tunnel, an analog computing device that measured the solutions to the associated nonlinear partial differential equations.

> Thus it was to a considerable extent a somewhat recondite form of computation which provided, and is still providing, the decisive mathematical ideas in the field of fluid dynamics. It is an analogy (i.e., measurement) method to be sure. It seems clear, however, that digital...devices have more flexibility and more accuracy, and could be made much faster under present conditions. We believe, therefore, that it is now time to concentrate on effecting the transition to such devices, and that this will increase the power of the approach in question to an unprecedented extent. [Goldstine and von Neumann, 1946, 5]

In this way, von Neumann's method for solving nonlinear problems led him to computing and to a general method for the use of the computer as an aid to mathematical research. The computer was to be employed to find numerical solutions of special cases of analytically intractable problems. These results would then provide a heuristic guide to theorizing about analytical solutions. About this approach he was highly sanguine:

> Really efficient high-speed computing devices may, in the field of nonlinear differential equations as well as in many other fields which are difficult or entirely denied of access, provide us with those heuristic hints which are needed in all parts of mathematics for genuine progress. [Goldstine and von Neumann, 1946, 4]

Examples from algebraic number theory and statistics illustrate how von Neumann used this technique in pure mathematical research. At the suggestion of their Princeton colleague, Emil Artin, von Neumann and Goldstine programmed the Institute computer to examine certain properties of prime numbers less than 10,000 of the form $p \equiv 1 \pmod 3$ [von Neumann and Goldstine, 1953]. In 1846 the mathematician E. E. Kummer had sought to generalize to the cubic case the famous quadratic sum result of Gauss. Kummer showed that for all

$$p \equiv 1 \pmod 3, \quad x_p = 1 + 2 \sum_{\nu=1}^{(p-1)/2} \cos(2\pi\nu^3/p)$$

satisfies the cubic equations $f(x) = x^3 - 3px - pA = 0$, where A is uniquely determined by the requirements $4p = A^2 + 27B^2$, $A \equiv 1 \pmod 3$. Since $f(x) = 0$ has three roots for all such p, Kummer classified each solution according to whether it was the largest, intermediate, or smallest root of $f(x) = 0$. He conjectured that the asymptotic frequencies for these classes are $1/2$, $1/3$, and $1/6$, respectively. Hand calculations of the first 45 x_p yielded Kummer the densities .5333, .3111, .1556. Using 15 million multiplications von Neumann and Goldstine showed the densities for $p \equiv 1 \pmod 3$ from 7 to 9973 to be .4452, .3290, .2258. This was a strong indication that Kummer's conjecture was false—a result of significance to Artin in his study of ideals.

A topic of considerable interest to mathematical statisticians in the late 1940s was randomness and the generation of random number sequences. Using the ENIAC, by then located at Aberdeen Proving Grounds, N. Metropolis, G. Reitweisner, and von Neumann [1950] generated long sequences of digits in the expansions of π and e and studied their randomness. (They found the digits of π, but not those of e, to conform with predicted randomness.) In collaboration with Bryant Tuckerman, von Neumann [1955] continued this research on the Institute computer. They generated long continued fraction expansions (2000 partial quotients) of certain real algebraic numbers, such as $2^{1/3}$, to compare statistics of these expansions with known distributions of these statistics over random numbers. Von Neumann also addressed the more general question of our capacity to produce enough random sequences of digits, given that statistical investigations on the computer consume them in great quantity. None of these statistical studies resulted directly in hard theorems, but they did provide mathematical researchers with insights in their exploration of randomness.

9.   THE MATHEMATICAL FOUNDATIONS OF COMPUTER SCIENCE

In addition to work on numerical methods for the computer and heuristic uses of the computer in mathematical research, von Neumann and his associates contributed to the mathematical foundations of computer science. This latter work was focused in three areas: the logical design of the computer; the planning and coding of problems; and the theory of automata. In each of these areas they brought mathematical methods and results to bear in the organization and advancement of what previously had been loosely organized arts. Their contributions were the first steps to providing the theoretical underpinnings needed for a science of computing.

Work on logical design, which had begun in the 1945 "Draft Report" on EDVAC, was continued in a 1946 paper by Arthur Burks, Goldstine, and von Neumann, entitled "Preliminary Discussion of the Logical Design of an Electronic Computing Instrument." While the purpose was to set forth the design of the Institute computer, the authors chose to treat the logical design issues more generally. Advantages and disadvantages of existing and prospective computer technologies were compared. As a result, for some years their work served as the principal source on logical design of stored-program computers, and a number of computers with similar architecture were constructed in addition to those originally planned by the government. Even today most computers incorporate the general principles of the so-called "von Neumann architecture" described in these reports [Rosin, 1983].

The "Draft Report" focuses on the logical character of computing devices with only passing reference to engineering details. Building on the logical framework established there, the first IAS report moved on to a detailed discussion of design issues. The authors compared the advantages and disadvantages of trigger circuits, gas tubes, and electromechanical relays, before selecting the cathode ray tube as the best available means of storing information. They described the architecture of the adder and the multiplier. The report adopted a systematic, scientific approach with detailed mathematical analysis. It analyzed the storage requirements for solving different kinds of mathematical problems, such as total and partial differential equations. In determining the specifications for an arithmetic unit, for example, it carried out a lengthy probabilistic argument showing that the sum of binary words, each of length n, has an average largest carry sequence not in excess of $\log_2 n$. It examined the mathematical properties of approximation, provided a detailed analysis of the use of the binary system in the computer, and gave an axiomatic presentation of control operations.

The "Draft Report" did not address programming and coding issues, even though von Neumann was experienced in preparing and coding problems for the ENIAC. During 1947 and 1948, he and Goldstine prepared a companion study that analyzed the general problems inherent in coding and programming. The result was a three part report, "Planning and Coding Problems for an Electronic Computing Instrument" [Goldstine and von Neumann, 1947b; 1948a; 1948b].

They identified two stages in the programming of a problem for the computer. First, one must decide which of those sequences of finitary operations that can be carried out by the computer will effect the appropriate computation. To aid this effort they developed a geometric system for displaying the logical sequence of operations that must flow forth to enact the computation. This technique was refined into a useful and sophisticated tool, which they called "flow diagramming." It is essentially a sophisticated version of our modern system of flow charts (see Figure 1). They noted the similarity of their flow analysis to work in mathematical logic; indeed, their flow diagramming notation plays an analogous role to the notational scheme in logic introduced by Frege in his *Begriffschrift* [1879].

The second stage is the "static coding," i.e., the writing of a set of rules, analogous to the description rules for Turing machines, that describe the internal working of the machine. These rules are "static" in that: each one carries out a fixed operation; they are applied one at a time; and any given rule takes effect only when the machine enters a given, predetermined internal configuration.

Four general principles were followed in developing coding procedures: (1) simplicity and reliability of the engineering solutions required by the coding; (2) simplicity, compactness, and completeness of the code; (3) ease and speed of translation into the code of human language and ease of finding and correcting errors; and (4) efficiency of the code in allowing the machine to work near its full intrinsic speed [Goldstine and von Neumann, 1948a, 81].

The report presented the flow diagrams and the static coding for a series of progressively more complex types of common mathematical problems: how to evaluate arithmetic expressions of the form $(au^2 + bu + c)/(du + e)$, given a value for u; how to convert

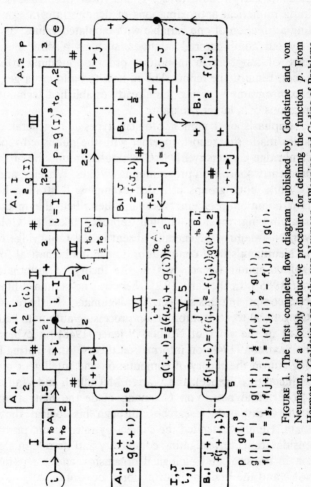

FIGURE 1. The first complete flow diagram published by Goldstine and von Neumann, of a doubly inductive procedure for defining the function p. From Herman H. Goldstine and John von Neumann, "Planning and Coding of Problems for an Electronic Computing Instrument," Report, Institute for Advanced Study, Princeton, NJ, 1947, Part II, Volume I, p. 18.

between the binary and decimal systems; how to carry out double precision arithmetic (in twice as many digits as the hardware was designed to carry); how to solve analytical problems, such as obtaining numerical approximations of definite integrals or interpolating a function of one variable with tabulated values; and how to carry out combinatorial methods, such as placing a random sequence of numbers in monotone order or meshing two such sequences. Finally, the report showed how to build up larger, more complex programs by using the already established techniques in subprograms of a larger program.

The emphasis on common problem types and general coding principles made this report, together with the published report of the Cambridge University EDSAC computer group, one of the two standard early sources on programming [Wilkes, Wheeler, and Gill, 1951]. While both groups provided examples illustrating how to program common problem types and described the method of subprogramming, the foundational approach taken by Goldstine and von Neumann was generally absent in the Cambridge report.

Von Neumann's final contribution to the mathematical foundations of computer science was in the theory of automata (see [Aspray, 1985]). In the early 1950s there was no standard literature or topics comprising this subject. Von Neumann clearly recognized that the general topic of information processing automata had been addressed in the work of Norbert Wiener [1948] on cybernetics, Claude Shannon [1948] on information theory, Alan Turing [1937; 1950; 1970] on the theory of computation and intelligent machinery (artificial intelligence), and Warren McCulloch and Walter Pitts [1943] on neural nets. Von Neumann made the first attempt to unify these disparate researches, through his logical theory of automata. He also extended the subject by introducing probabilistic considerations, investigating complexity and its role in self-replication in physical and biological organisms, and comparing the computer and the brain as information processors.

His logical theory of automata aimed to use mathematical logic to develop a coherent theory in which computers and brains were considered to be instances of a single type of object that processes information. He did not intend to study particular mechanical or physiological features of the devices that carry out the processing, but only the "general theoretical regularities that may be detectable in the complex synthesis of the organism in question," [von Neumann, 1951, 289–290]. To that end, individual neurons and computer elements were treated as "black boxes," totally unknown except for their behavior, which was expressed by axioms.

Von Neumann made few contributions to the logical theory of automata beyond explicitly drawing out the connections and fundamental results in the work of others. For example, he was the first to observe that to each Turing machine there corresponds a McColloch and Pitts neural net that is functionally equivalent in the sense that they both carry out the same information processing activity. Since the Turing machine represents the artificial physical world and the neural net the biological world, von Neumann reasoned that he had established a fundamental connection. He also identified some limitations of formal logics as they apply to the study of practical information processing, e.g., that formal logics are insensitive to the varying finite lengths of a computation or to the effect that probability of error at a given step in a computation has upon the probability of error in the complete process.

The second of these concerns was the principal subject of von Neumann's probabilistic theory of automata [von Neumann, 1956]. He hoped that it would provide the tools to study how automata with unreliable components are able, or can be made, to function reliably. Given the probability of error of a black box, he developed rigorous methods to calculate the probability of error in an automaton comprised of them. This led him to introduce the technique of *multiplexing* to enhance the reliability in an automaton with unreliable components. In this technique each fundamental operation is processed and transmitted by N identical black boxes, the size of N determined by the overall level of reliability required of the automaton. He calculated that, given certain assumptions about the reliability and switching frequency of vacuum tubes, an electronic computer of his day would achieve mean free path to error of 8 hours if multiplexed 17,500 times. He also noted the fundamental connection between his work in probabilistic theory, Shannon's research on noise in an information channel [1949], and that of Ludwig Boltzmann [1896; 1898], R. V. L.

Hartley [1928], and Leo Szilard [1929] on statistical mechanics. Finally, he suggested that intuitionistic logic might serve as the foundation for the theory of information processing automata with unreliable components.

Von Neumann's study of complexity and self-replication in information processing automata was prompted by the observation that the two prime examples, the computer and the brain, are highly "complicated" in the sense that they have many logical switches and can perform many different tasks. Admittedly, what he meant by "complication" he could not make precise:

> There is a concept which will be quite useful here, of which we have a certain intuitive idea, but which is vague, unscientific, and imperfect. This concept clearly belongs to the subject of information, and quasi-thermodynamical considerations are relevant to it. I know no adequate name for it, but it is best described by calling it "complication." It is effectivity in complication, or the potentiality to do things. I am not thinking about how involved the object is, but how involved its purposive operations are. In this sense, an object is of the highest degree of complexity if it can do very difficult and involved things.   [von Neumann, 1966, 78]

One complicated activity von Neumann considered was the ability to self-replicate. Certain biological organisms, he noted, are able to produce off-spring of the same level of complication or even to evolve into higher biological forms; while other automata, e.g., machine tools, are able only to produce automata less complicated than themselves. He speculated that this is a fundamental property, one related to evolution and to intelligence in machinery.

> There is thus this completely decisive property of complexity, that there exists a critical size below which the process of synthesis is degenerative, but above which the phenomenon of synthesis, if properly arranged, can become explosive, in other words, where synthesis of automata can proceed in such a manner that each automaton will produce other automata which are more complex and of higher potentialities than itself.   [von Neumann, 1966, 80]

Von Neumann first investigated the self-replicative properties of the universal Turing machine (the one able to reproduce the behavior of all Turing machines). Then, in 1952 and 1953, he designed one physical and three mathematical models of self-replicating automata [von Neumann, 1966]. The mathematical models are cellular automata: infinite, two-dimensional arrays of square cells. Each cell assumes one of a number (between 29 and infinity, depending on the model) of unexcitable, excitable, or excited internal states. The internal state of a cell could be modified either by excitation transmitted from contiguous cells or by "fatigue." "Intelligence" was represented by the patterns represented in the configuration of excited cells, and replication was considered to have occurred if an "intelligence" configuration was copied from one region of the cellular automaton to another.

Von Neumann's last contribution to automata theory was a direct comparison of the brain and computer as information processors [von Neumann, 1958]. He began by comparing their basic logical switches—the neuron and the vacuum tube, respectively—for speed, energy consumption, size, efficiency, and number required in a system. Next he compared them as total information processing systems, evaluating the number of operations required to carry out a task, the level of precision and reliability, the method of handling errors, and the means of storage, control, input, and output. His hope was that such a study of natural systems would provide object lessons for the design of artificial information processing systems (computers).

10. CONCLUSIONS

The Second World War stimulated great changes in calculating equipment and its use. Many engineering and scientific problems—including cryptanalysis, ballistics table-making, bomb and aircraft design, and signal processing—required advances in calculating technology. These needs stimulated the development of both analog and digital devices, most importantly the electronic, stored-program, digital computer. Two lines of digital equipment emerged during the war and continued thereafter: computers for processing large numbers of arithmetic operations, and computers for processing symbols and only a few arithmetic operations.

The Institute for Advanced Study Computer, a machine of the number processing variety designed for fundamental scientific computation, was built through the vision and efforts of John von Neumann. His plan to build a computer at the Institute was not

supported by the other members of the faculty, who had neither a
need for, nor an interest in, such equipment. Von Neumann was
able to attract a group of able mathematicians and physical scientists to the Institute to use the computer, including many of the
early leaders in modern numerical analysis and scientific computation. In the late 1950s most of the mathematical interest in the
computer resulted from applied mathematicians working in areas
like hydrodynamics, although there was an occasional "pure"
mathematical use like number theory.

The "inner economy" of the computer as contrasted with earlier
calculating technology made storage expensive and multiplication
cheap; this required new numerical methods attuned to this economy, for example, to solve systems of linear equations, invert
matrices, and evaluate definite integrals. The renewed interest,
during and after the war, in applications involving partial differential equations and the ability of the computer to carry out many
more arithmetic operations in a numerical approximation, brought
attention to the question of the stability of numerical methods. The
complexity of the computer also led mathematicians, especially
logicians, to turn attention to the foundations of computing science: logical design and switching theory, computability, automata
theory, and information theory. These were the beginnings of the
much closer tie between mathematics and computing that exists
today.

Acknowledgement.   I would like to thank Garrett Birkhoff, Herman Goldstine, and
several anonymous referees for their criticisms of an earlier draft of this article.

BIBLIOGRAPHY

Aspray, William, 1985, "The Scientific Conceptualization of Information: A
Survey," *Annals of the History of Computing* 7, 117–140.

Bargmann, Valentine, Deane Montgomery, and John von Neumann, 1946, "Solution of Linear Systems of High Order," U.S. Navy Bureau of Ordnance Report
under Contract Nord - 9596.

Birkhoff, Garrett, Forthcoming, *History of Numerical Fluid Dynamics.*

Boltzmann, Ludwig, 1896, *Vorlesungen über Gastheorie*, Pt. I, J. A. Barth, Leipzig.

———, 1898, *Vorlesungen über Gastheorie*, Pt. II, J. A. Barth, Leipzig.

Courant, Richard, Karl Friedrichs, and Hans Lewy, 1928, "Über die partiellen
Differenzengleichungen der mathematischen Physik," *Math. Annalen* 100, 32–74.

Frege, G., 1967, "Begriffsschrift," in *From Frege to Gödel*, (ed.) Jean van Heijenoort, Harvard University Press, Cambridge, Mass., pp. 1–82.

Goldstine, Herman H., 1972, *The Computer from Pascal to von Neumann*, Princeton University Press, Princeton.

Goldstine, Herman H. and John von Neumann, 1946, "On the Principles of Large Scale Computing Machines." Unpublished, reprinted in Abraham Taub (ed.), *John von Neumann Collected Works V*, Macmillan, New York, 1963, pp. 1–33.

_____, 1947a, "Numerical Inverting of Matrices of High Order," *Bulletin of the American Mathematical Society* 53, 1021–1099.

_____, 1947b, "Planning and Coding of Problems for an Electronic Computing Instrument," Part II, Vol. 1, report prepared for U.S. Army Ordnance Department.

_____, 1948a, "Planning and Coding of Problems for an Electronic Computing Instrument," Part II, Vol. 2, report prepared for U.S. Army Ordnance Department.

_____, 1948b, "Planning and Coding of Problems for an Electronic Computing Instrument," Part II, Vol. 3, report prepared for U.S. Army Ordnance Department.

_____, 1951, "Numerical Inverting of Matrices of High Order, II," *Proceedings of the American Mathematical Society* 2, 188–202.

Hartley, R. V. L., 1928, "Transmission of Information," *Bell System Technical Journal* 7, 535–553.

Hyman, M. A., 1947 (October 16), "Stability of Finite Difference Representation," Naval Ordnance Laboratory, Mechanics Division Technical Note SB/Tm-18, 2–11.

McCulloch, Warren S. and Walter Pitts, 1943, "A Logical Calculus of the Ideas Immanent in Nervous Activity," *Bulletin of Mathematical Biophysics* 5, 115–133.

Metropolis, N. C., G. Reitweisner, and John von Neumann, 1950, "Statistical Treatment of First 2000 Decimal Digits of e and π calculated on the ENIAC," *Mathematical Tables and Other Aids to Computation* 4, 109–111.

O'Brien, George C., Morton A. Hyman, and Sidney Kaplan, 1950, "A Study of the Numerical Solution of Partial Differential Equations," *Journal of Math. Physics*, 29, 223–239.

Richtmyer, R. D., 1957, *Difference Methods for Initial-Value Problems*, Interscience, New York.

Rosin, R. F., 1983, "Von Neumann Machine," in Anthony Ralston (ed.), *Encyclopedia of Computer Science and Engineering*, 2nd ed., pp. 1565–66.

Shannon, Claude, 1948, "The Mathematical Theory of Communication," *Bell System Technical Journal*, 27, 379–423 and 623–656.

_____, 1949, "Communication in the Presence of Noise," *Proceedings of the Institute for Radio Engineers*, 37, 10–21.

Southwell, R. V., 1946, *Relaxation Methods in Theoretical Physics*, Clarendon Press, Oxford.

Szilard, Leo, 1929, "Über die Entropieverminderung in einem thermodynamischen System bei Eingriffen intelligenter Wesen," *Zeitschrift fur Physik* 53, 840–856.

Turing, Alan, 1937, "On Computable Numbers, with an Application to the Entscheidungsproblem," *Proceedings of the London Mathematical Society*, Series 2, 42, 230–265.

Turing, Alan, 1950, "Computing Machinery and Intelligence," *Mind*, n.s., 59, 433–460.

———, 1970, "Intelligent Machinery," in B. Meltzer and D. Michie (eds.), *Machine Intelligence*, Vol. V, American Elsevier, Edinburgh, pp. 3–26.

Von Neumann, John, 1945, "First Draft of a Report on the EDVAC," report prepared under U.S. Army Ordnance Contract 2-670-ORD-4926. Reprinted in Nancy Stern, *From ENIAC to UNIVAC*, Digital Press, Bedford, MA, 1981, pp. 177–246.

———, 1948a (June 22–July 6), "First Report on the Numerical Calculation of Flow Problems," unpublished report. Reprinted in Abraham Taub (ed.), *John von Neumann Collected Works* V, Macmillan, New York, 1963, pp. 664–712 with an editorial note.

———, 1948b (July 25–August 22), "Second Report on the Numerical Calculation of Flow Problems." Reprinted in Abraham Taub (ed.), *John von Neumann Collected Works* V, Macmillan, New York, 1963, pp. 713–750.

———, 1951, "General and Logical Theory of Automata," in *Cerebral Mechanisms in Behavior—The Hixon Symposium*. Reprinted in Abraham Taub (ed.), *John von Neumann Collected Works* V, Macmillan, New York, 1963, pp. 288–328.

———, 1956, "Probabilistic Logics and the Synthesis of Reliable Organisms from Unreliable Components," in C. E. Shannon and J. McCarthy (eds.), *Automata Studies*, Princeton University Press, Princeton.

———, 1958, *The Computer and the Brain*, Yale University Press, New Haven.

———, 1966, *Theory of Self-Reproducing Automata*, edited and completed by Arthur W. Burks, University of Illinois Press, Urbana.

Von Neumann, John and Herman H. Goldstine, 1953, "A Numerical Study of a Conjecture of Kummer," *Mathematical Tables and Other Aids to Computation* VII, No. 42, 133–134.

Von Neumann, John and R. D. Richtmyer, 1947 (December 25), "On Solution of Partial Differential Equations of Parabolic Type," Technical Report, Los Alamos Scientific Laboratory, LA-657.

Von Neumann, John and Bryant Tuckerman, 1955, "Continued Fraction Expansion of $2^{1/3}$," *Mathematical Tables and Other Aids to Computation* 9, pp. 23–24.

Wiener, Norbert, 1948, *Cybernetics: or Control and Communication in the Animal and the Machine*, Technology Press, Cambridge.

Wilkes, Maurice V., David J. Wheeler, and Stanley Gill, 1951, *The Preparation of Programs for an Electronic Digital Computer*, Addison-Wesley, Reading, Mass. Reprinted in the Charles Babbage Institute Reprint Series for the History of Computing, Tomash Publishers, Los Angeles and San Francisco, 1982.

LUDWIG BIEBERBACH AND "DEUTSCHE MATHEMATIK"

Herbert Mehrtens[1]

> Of all fields of intellectual culture thus far only mathematics seemed to keep its neutrality in the furnace of the revolution. A remarkable lecture, held on Easter-Tuesday by Professor Bieberbach... appears to indicate that the doctrine of blood and race reaches out even here and subordinates the most abstract of all sciences to the total state. [Anon., 1934]

These are the first sentences of a detailed and ironic newspaper report under the heading "New Mathematics." The news of Bieberbach's ideas spread quickly and prompted sharp reactions from abroad. In a letter to the editor of *Nature*, the British mathematician G. H. Hardy related the contents of the lecture and drew the conclusion:

> It is not reasonable to criticize too closely the utterances, even of men of science, in times of intense political or national excitement. There are many of us, many Englishmen and many Germans, who said things during the war which we scarcely meant and are sorry to remember now. Anxiety for one's own position, dread of falling behind the rising torment of folly, determination at all costs not to be outdone, may be natural if not particularly heroic excuses. Prof. Bieberbach's reputation excludes such explanations of his utterances; and I find myself driven to the more uncharitable conclusion that he really believes them true. [Hardy, 1934]

195

All evidence suggests that Bieberbach in fact believed in what he said. Recently his name has been cited frequently because of the proof of the so-called "Bieberbach conjecture." In this paper, however, we are concerned with the second, political, Bieberbach conjecture which stated, in the first place, that two basic styles of doing mathematics existed and that the more concrete, reality-oriented, intuitive style—as represented by Felix Klein—was of higher value than that of modern, formalist mathematics; moreover, these styles were related to types of personality and ultimately to human races, so that the better style—because it was characteristic for the Germans—could be installed with the help of National Socialism as a norm for mathematical work in Germany.

Bieberbach's political conjecture is epitomized in the name of the journal he founded in 1936, *Deutsche Mathematik*. Although this journal survived until 1944, when the last part of the 1942 volume was published, it was a disproof of Bieberbach's conjecture. For Bieberbach could not convince his fellow mathematicians in Germany to believe in his conceptions of a sound "German" mathematics, nor did he find sufficient political backing to force them to accept his ideas.

Bieberbach's *Deutsche Mathematik*, like the *Deutsche Physik* of Philip Lenard and Johannes Stark, has since become a symbol for the inadmissable intrusion of politics into science. How such a thing could happen and what the conditions were that determined its fate has as yet not been sufficiently analyzed.[2]

In this paper I shall briefly describe Bieberbach's early career and give a detailed study of his ideas about mathematical styles and his nationalist and anti-Göttingen activities during the twenties. These set the precedent and motivation for his later Nazi activities and explain in what sense he was "subjectively honest." The second part of the paper deals with Bieberbach's activities during the time of Nazi rule; here his theories of styles, types, and races will be discussed. Bieberbach's story reveals an inextricably interlocked pattern of scholarly values, political positions, institutional interests, and personal relations. Such patterns are by no means uncommon, not even in "the most abstract of all sciences." And, as Hardy wrote, "in times of intense political or national excitement" they can lead to drastic consequences.

Intolerable and inexcusable as the consequences in the present case are, they do not defy historical explanation. The patterns revealed in Bieberbach's case are not altogether unfamiliar in fairly "normal" conflicts in scientific disciplines. It should be noted, however, that the explanations offered here are only partial ones. Archival sources are insufficient, for example, to evaluate in greater detail the role of pressures from private life or to give a full picture of the personality of our protagonist. Nor has an attempt been made to give a description and analysis of his mathematical work.*

EARLY CAREER

Ludwig Bieberbach was born on December 4, 1886 in Goddelau, a small town south of Frankfurt.[3] His father was the director of the local state mental institution, where his grandfather had also worked as a psychiatrist. While serving his year of military service (1905–6) in Heidelberg, Bieberbach continued to follow his early keen interest in mathematics by taking university courses. His interest in algebraic forms led him to Göttingen in 1906. His teachers were Paul Koebe, at the time a *Privatdozent*, and Felix Klein, who deeply influenced Bieberbach in his views on mathematics. Work with these two led to his doctoral dissertation on automorphic functions [Bieberbach, 1910]. A report in Klein's seminar on Arthur Schoenflies' work on crystallographic groups revived his algebraic interests. His *Habilitation* [Bieberbach, 1911–12] on groups of Euclidean motions was a first important step to the solution of Hilbert's eighteenth problem [Alexandrov, 1971].

When Ernst Zermelo received an appointment at Zurich in 1910 Bieberbach followed him and became a *Privatdozent* there, but left before the year was over for Königsberg, where Schoenflies had arranged a teaching assignment for him. In 1913 he accepted a full professorship in Basel. Meanwhile Schoenflies was in Frankfurt, where he played a role in establishing the new university at which Bieberbach obtained a chair in 1915. Being unfit for military service Bieberbach remained in his position at the university throughout World War I. In 1916 he published his well-known conjecture on the coefficients of *schlicht* functions, providing a proof for a special case. Bieberbach's career progressed rapidly, and in 1921 he was chosen to be the successor of Constantin Carathéodory at Berlin University, one of the two leading institutes of mathematics at that time.

Bieberbach's nomination at Berlin sheds some light on his ranking among German mathematicians [Bieberbach, AS 5b]. In 1917 he had been proposed for an extraordinary professorship at Berlin, but the decision was postponed. In the same year Georg Frobenius, who occupied one of the three chairs in mathematics at Berlin University, died. Issai Schur, Carathéodory, Herman Weyl, Erich Hecke, Gustav Herglotz, and Adolf Kneser were, in this order, proposed as possible successors (Hilbert did not appear on the list because he had already twice rejected offers from Berlin). Carathéodory was appointed but left Berlin by 1919. A new list naming L. E. J. Brouwer, Herglotz, and Weyl was prepared. Later Hecke's name was added but all of these declined the offered position. The Berlin faculty had to start again from scratch, this time naming Wilhelm Blaschke, Bieberbach, Gerhard Hessenberg, and Ernst Steinitz. The quality of the research of Blaschke and Bieberbach was rated equal; Blaschke was preferred because his interests were most suited to Berlin. The statement prepared for the ministry mentioned Bieberbach's "excellent work" on geometrical and algebraical groups and his more recent "beautiful results" in the geometric theory of complex functions:

> Even if his presentation sometimes lacks the desirable precision, this is by far outweighed by the vividness of his scholarly initiative and the broad-minded and deep approach of his research.

Another document attesting to Bieberbach's reputation is the proposal for his election, which took place in 1924, to the Prussian Academy of Sciences [Bieberbach, AS 6]. The proposal describes his achievements in some detail and concludes:

> Bieberbach is a versatile researcher whose wealth of ideas leads quickly to positive and valuable results.

In accepting election to the academy Bieberbach delivered an introductory address [1924a], in which he began by explaining that he considered himself to be a "geometric thinker":

> whether I have worked in the theory of complex functions or in group theory, in analysis or in arithmetic, my reasoning always had its base in a figurative geometric conception.

After this general characterization he briefly described his major mathematical achievements, mentioning the textbooks he had written as an additional service he had performed. Moreover, he added, his "sense of duty" had led him to undertake "activities for the external life of my science." Here he cited his position in the German mathematicians' association (*DMV*)—where he had served since 1921 as permanent secretary and editor of its journal, the *Jahresbericht*—and his activities on the mathematics committee of the recently founded Emergency Association of German Science (*Notgemeinschaft*) (cf. [Tobies, 1981]). Also, since 1920 Bieberbach had been on the editorial board of *Mathematische Annalen*.

Bieberbach had, in fact, been very productive, doing research in a wide range of topics. His numerous book reviews and several expository articles show a strong interest in the history and philosophy of mathematics and in its applications. Bieberbach also wrote on numerical analysis [1921a] and on nomography [1922; 1924b]. By 1924 he had published textbooks on differential and integral calculus [1917a; 1918], conformal mapping [1915], the theory of complex functions [1921b; 1922a], and on differential equations [1923]. The latter book was published as the sixth volume of Springer's *Grundlehren* series, edited by Richard Courant. Although some of Bieberbach's writings was considered a little sloppy (mathematically) these books were very successful and went through numerous printings.

Documents I have studied in the preparation of this paper, as well as conversations I have had with former students and colleagues of Bieberbach, suggest that he was, both physically and intellectually, a very lively man, sometimes expressive to a comical degree. He was a somewhat muddled lecturer but very friendly and helpful to his students, a little conceited but not arrogant, all in all a slightly eccentric, but sympathetic character. A typical anecdote comes from a letter Albert Einstein wrote in 1919 to Max Born's wife, Hedwig:

> Herr Bieberbach's love and devotion for himself and for his muse is quite priceless. May God keep it with him, this is the best way to live.

Max Born recalled that at Frankfurt University there was a book containing brief autobiographical statements by the faculty members.[4] Born showed the book to his wife who found young Bieberbach's entry "rather funny" and "exuberant with vanity" [Einstein and Born, 1972, 24–26].

Bieberbach's later turn to Nazism apparently surprised his students and most of his colleagues. Although some psychological anomalies have been suggested nobody has offered a plausible explanation. Since there is insufficient evidence for psychological assessment, we must examine carefully those events and ideas which provide at least a partial explanation of later developments.

One of the problems related to Bieberbach's turn to the Nazis is what he thought about Jews. Although he had Jewish friends and students, he still willingly served the anti-Semitic regime. One incident earlier in his career might be noted in this connection. After accepting the offer of a chair in Berlin, Bieberbach became very actively involved in selecting his successor in Frankfurt. In letters concerning this matter he openly asked possible candidates if they were Jewish [Bieberbach, AS 3c], and in his correspondence with Blaschke, Bieberbach [AS 3d] discussed the "advantage" of being Jewish in obtaining a position at Frankfurt. His own bias then was against foreigners and not against Jews [Klein, AS 1]. The University of Frankfurt had been founded with private donations from wealthy and predominantly Jewish Frankfurt citizens, hence the unique pro-Jewish policy at the University. This illustrates how Jews in German universities and in scientific occupations were, in fact, a discernible, separate group who found anti-Semitism in some places, occasionally received privileges—as in Frankfurt—and encountered problems and reservations almost everywhere. Bieberbach perceived—certainly not only in this case—that Jews were objects of politics. The idea that they were racially different was certainly in the air and was not unfamiliar to Bieberbach. He knew and later quoted (below) Felix Klein's remark on the "logical sense" of the "Hebrew race." Similarly, Sommerfeld, in a letter (1907) had written of the "abstractly conceptual style of the Semite" [Kleinert, 1985]. Prior to 1933, no such utterances by Bieberbach are known. He was, however, accustomed to hearing such ideas expressed, and did not associate them with anti-Semitism, as the cases of Klein and Sommerfeld show (cf. [Rowe, 1984] on Klein).

"IDEALS OF SCIENCE" IN MATHEMATICS

Bieberbach described himself as a "geometrical thinker," an assessment confirmed by the statement of the department quoted above. At the same time he had a strong interest in the unity of mathematics, not only in his own broad areas of mathematical research but also as a matter of editorial policy for the *Jahresbericht der DMV*:

> We want to help propagate the awareness of the unity of mathematics, we want to keep everybody informed about the most important advances and thus help to restrain the rising danger of suffocation in narrow specialism which ultimately must lead to the end of scholarship.   [Bieberbach, 1922b][5]

The unity of mathematics as well as the value and necessity of geometrical intuition are recurring topics in Bieberbach's publications.

In his inaugural address—his first publication on the philosophy of mathematics—Bieberbach [1914] took a clear formalist standpoint. He depicted "modern mathematics" as making a sharp distinction between applied mathematics, which is concerned with the possibilities of a conceptual grasp of the real world, and pure mathematics as an "abstract doctrine of relations" [Bieberbach, 1914, 898]. After discussing formalism and intuitionism he concluded that formalism was the only point of view "that does justice to the factual state of mathematical science... . We do not want to create knowledge of the truth but merely methods to gain knowledge... . The truth of mathematics rests solely in its logical correctness and consistency" [Bieberbach, 1914, 901]. Yet Bieberbach remained, at this time, rather vague on many points. On the whole he took the position of a working mathematician: he was willing to accept the unsolvability of basic foundational problems of mathematics and called for axiomatic systems adequate to the subject matter as it had historically developed to its present state. There are no comments in this article on the relative heuristic values of intuition and logic.

In a book review Bieberbach [1917b] defended axiomatics against the assertion that it would turn mathematics into a mere game with symbols. He reiterated the "sharp distinction" between axiomatics as the "logical study of conceptual schemes" and the study of applicability to real objects. The preface to his books on integral and differential calculus [Bieberbach, 1917a; 1918], however, reveals a somewhat different position. Here Bieberbach explained that he had allowed for applied problems in the book, and went on:

> To me it seems to be not only historically a mistake to tear these things away from 'pure' mathematics in order to turn them, along with other matters, into a new field called 'applied mathematics' which some people would like to wedge between pure mathematics and its fields of application. This unintentionally destroys the living spirit of outside stimulations which created mathematics and to which it constantly and abundantly owes new life.

He added that he did not question the need of a comprehensive treatment of problems of applicability but this works only on the basis of theory and not without or even against theory:

> I have always attempted to show how the concepts of mathematics are basically conceptual renderings and thus stylized forms of contents of imagination which in themselves are unfit as basis of logical operations.

In an article written on the occasion of Felix Klein's seventieth birthday we find the theme Bieberbach was to repeat in his later publications on the "ideals" and "styles" of mathematics:

> The tendency towards formalism must not cause us to forget the flesh and blood over the logical skeleton.... The formalism of university education in mathematics has forgotten about applications and the cultural meaning of mathematics. [Bieberbach, 1919]

The quoted passages from 1914 to 1919 show Bieberbach's shift to an antiformalist position, that was later to become even more radical. There was, to be sure, a certain consistency in his approach to philosophical problems from the standpoint of the working mathematician. There is no indication that he ever studied or taught mathematical logic or foundations proper. In Frankfurt he announced (as German mathematicians regularly did at that time) a series of lectures of a philosophical kind to be given in 1919 and 1920: the topics were, "On the nature of mathematics," "History of modern mathematics," and "Problems of mathematical research." There is no record on the content of the courses, although it is likely they were—like his publications—presentations of a historical and pragmatic, so to speak 'private,' philosophy of mathematics, in which he explained to students what he called the "cultural meaning" and the "flesh and bones" of mathematics.

The motivations for Bieberbach's shift from his 1914 argument in favor of formalism to the anti-formalist warnings of 1919 are difficult to understand. In 1914 he used the argument which in the 1920s reappeared in Hilbert's defense of formalism, namely that intuitionism cannot meet the necessities of the business of mathematics (*Betrieb der Wissenschaft*). Later he saw the same business as "suffocating" under the pressure of specialization and the formalist tendency to deprive mathematics of meaning and common sense. Obviously this does not imply a new choice of a technical basis for mathematics; it is rather a massive shift in values.

One of the main causes of this reevaluation must have been World War I. Two examples illustrate Bieberbach's nationalism at this time. In the lecture of 1914 Bieberbach had remarked, "it is no wonder that the main protagonists of intuitionism are at home in France" [Bieberbach, 1914, 900], implying that French mathematicians stubbornly defended old beliefs, like the mathematicians of the eighteenth century who were unable to imagine that the problem of parallels could find its solution in another geometry than that of Euclid. The Germans, on the other hand, were able to progress into new areas of thinking. In the wake of World War I such nationalist undertones were quite common in the writings of German scholars. Towards the end of the war, after the central powers had already started their last and initially successful offensive, Bieberbach dated the preface to his book on the integral calculus [1918, iv] to the "*siegerfüllten*" (filled with victory) spring of 1918. His unrealistic hope, at this late date, for a German victory was by no means uncommon at that time.

But, by the end of 1918, the yearned for victory had turned into a deeply disillusioning and humiliating defeat. For some the disillusionment had started much earlier, as the brittle facade of social harmony, erected to hide the tense realities of Wilhelmian Germany, began to erode. The time had come for a reconstruction or a radically new construction of social and intellectual worlds, including the world of mathematics. I have argued elsewhere ([Mehrtens, 1984], cf. also [Mehrtens, 1982]) that the so-called "crisis" in the foundations of mathematics was indeed a social crisis embedded in this general environment. The technical core of the two or three foundational programs had certainly generated a debate, but not a "crisis." What turned this debate into a crisis were individual and professional aims; the value system and the basic epistemological beliefs in mathematics connected with aims; and values and beliefs of a more general nature.

Additional research on the role of the war in changing attitudes in science is necessary, and not only in Bieberbach's case. We may note that in 1917 L. E. J. Brouwer returned to foundational studies, and that during the war Hermann Weyl, who declared Brouwer to be the "revolution" in mathematics [Weyl, 1921], had turned from an outright formalist position to intuitionism. Similarly, both Lenard and Stark, the leading proponents of a National Socialist *Deutsche Physik*, appear to have arrived at their mixture of political and epistemological beliefs through the experience of the war (cf. [Beyerchen, 1977, Chapters 5, 6]).

Bieberbach was part of this contemporary ambience in which questions of philosophy, professional policy, school reform, finance, and the publication system, of fragmentation and integration of mathematical sciences, and of international relations were all closely interrelated. As we have seen, Bieberbach was actively involved in the politics and organization of his discipline, and in the early 1920s he appears to have been fairly successful and uncontroversial in this role. Serious conflicts occurred only during the second half of the decade. Before considering those developments in the next section, Bieberbach's nontechnical publications, written before 1933, will be analyzed.

On the occasion of Hilbert's birthday Bieberbach prepared an article [1922d] in which he failed to mention the latter's foundational work. A short time later, however, he was speaking of the "late rampant growth of Hilbert's axiomatics in the field of physics" [Bieberbach, 1924c, 726]. In 1925 Bieberbach published an article, "On the Development of Non-Euclidean Geometry in the 19th Century," whose underlying theme is the interrelation of mathematical truth, scientific empirics, and *Anschauung* (which I shall not attempt to translate), a key concept in Bieberbach's ideas. This notion touches on visual perception, geometrical intuition or visualization, as well as "intuition," in the sense of Brouwer or Kant. It remained sufficiently vague to have different meanings in different contexts, and it set a sharp contrast to formal operations and reasoning. Bieberbach held that the ideas of the founders of non-Euclidean geometry were rooted in experience and *Anschauung*. Here, as elsewhere, he used Gauss' geometrical representation of complex numbers as the example of how "ideal" or "imaginary" concepts achieve "citizenship" in mathematics through *Anschauung*. In the end he characterized Hilbert's foundational work as an attempt "to push back the role of intuition even further" [Bieberbach, 1925, 397], adding that this effort had met with opposition from intuitionists; although his leanings towards the latter position were implicit here he still did not take a clear stand.

By 1925, however, Bieberbach had established close relations with Brouwer, especially on political matters concerning the international boycott of German science (discussed below). In a remarkable lecture, given in 1926, Bieberbach openly sided with Brouwer and sharply attacked Hilbert's "ideal of science." The views stated in this lecture are related to two translations: one, made by Bieberbach himself, was of Frederigo Enriques' *History of Logic* [1927], published originally in Italian. The book consists of a history of logic and an extensive discussion of contemporary positions on the philosophy of logic and what is called the "logic of science," which is not mathematical logic. Enriques' treatment is characterized by a strong element of psychologism, a certain historicism, and, as a third element close to Bieberbach's ideas, an emphasis on the meaning of reality and imagery for science [Enriques, 1927, 209f]. Enriques called for a "logic of scientific systems" as a synthesis of the deductive logic of abstract concepts with the "logic of passion," as the movement of ideas on an emotional basis within the unity of thinking and reality [Enriques, 1927, 219f].

The other translation initiated by Bieberbach was of Pierre Boutroux's *L'idéal scientifique des mathématiciens* [1927] (cf. the preface). According to his own statements (below), Bieberbach was deeply influenced by Boutroux's book, adopting many of its ideas and even phrasings. Boutroux had maintained the existence of two conflicting "orientations of thought," described alternately as synthetic vs. analytic, intuitive vs. discursive, or as order of invention vs. order of proof. He postulated an "intrinsic" objectivity of mathematics, a realm of mathematical facts independent of constructions. The motives of mathematicians, he wrote, "have almost nothing in common with the algebraico-logical presentation" [Boutroux, 1927, 182] which is, instead, an

> edifice of symbols which we pile up to infinity—like a clever juggler would do, who finds joy in multiplying the difficulties of his exercises.   [Boutroux, 1927, 184]

Boutroux accepted intuitionist ideas, although in the sense of Poincaré rather than Brouwer, but nevertheless concluded:

> No theory which tends to conceive of the variety of mathematical thinking as a unity is capable of giving an appropriate picture of the present condition of analysis.   [Boutroux, 1927, 205]

In the closing pages of the book Boutroux discussed mathematics education, pleading that the completion of the anti-formalist reforms requires not only the cultivation of sensual intuition but also the training of intellectual intuition.

For his lecture of 1926 Bieberbach [AS 2b] simply adopted the title of Boutroux's book. The two "ideals of science" ("*Wissenschaftsideale*") are Hilbert's formalism on the one hand and *anschauliche Gegenständlichkeit*—the object related thinking of Felix Klein—on the other. The latter coincided with what Bieberbach had (in his Academy address) referred to as "geometrical thinking" being led by the *Anschauung*. He also used the term "style" for these two alleged modes of mathematical thought. This notion implies a combination of foundational philosophy, conceptions of objects and aims of mathematical research, methodology, and the manner of literary expression in mathematical writing.

The formalism attacked by Bieberbach is, in this sense, the style of "modern" mathematics, as exemplified, for example, in van der Waerden's well-known *Moderne Algebra* [1930]. The "modern" approach in algebra was much discussed in those years as was the "modern" rigorous manner of exposition, e.g., the "Landau-style" that Bieberbach was later to condemn as "un-German." Helmut Hasse's review [1929] of Edmund Landau's *Vorlesungen über Zahlentheorie* contains a lengthy discussion of the problems and merits of this style.

What was called "modern" in those years was the kind of mathematics which, after the turn of the century took the lead in the field of mathematical research and proved most productive. It was, in the sense of Lakatos [1968] the most progressive research program in mathematics. If the advent of this "modern" program may be interpreted as a "scientific revolution" in mathematics, then Bieberbach was on the side of the losers in this revolution. His description of formalism may well reflect his fears of being left behind by the establishment of new standards in assessing research —standards to which he either could not or did not wish to conform. Formalism, to Bieberbach, was seen as the "prevalent orientation," the "fashion of the day," whose proponents tried

to turn a method, which in its proper place can be successful, into the prototype for all of mathematics, the prototype for all that is allowed to be called mathematics. [Bieberbach, AS 2b, 17]

Obviously to define what is to be "allowed" in mathematics is a matter of power within the social system of the discipline, and Bieberbach's subsequent conflicts (discussed below) with Hilbert and Springer must be seen in the light of his conviction that there was an unjustified dominance of the formalist "ideal." Bieberbach left no stone unturned in his denouncement of formalism: it is "unsound," "deadening," "logical nonsense," "paradoxical over-estimation of logical form," and leads to the "removal of sense and meaning from mathematics." He added, more concretely, that

under the reign of formalism applied mathematicians have, so to speak, become extinct.... . What role, moreover, is this ideal to play in the schools and in the training of future teachers? How does one imagine the drill of children for such eccentricities? [Bieberbach, AS 2b, 19]

The intuitionism of Brouwer and Weyl, on the other hand, was to Bieberbach like "a breath of spring air," "the escape from a nightmare":

> Intuitionism attempts to account for the fact that human beings do mathematics, and to make allowance, systematically, for *anschauliche* concreteness everywhere in its constructions.   [Bieberbach, AS 2b, 21]

To a large extent the lecture of 1926 was a *laudatio* to Felix Klein, whose ideal, according to Bieberbach, was

> the development of *Anschauung*, the training, regimentation, and refinement of *Anschauung* and thinking.   [Bieberbach, AS 2b, 16]

In Klein's time, however, historical development (here Bieberbach repeated the ideas of his paper on non-Euclidean geometry) had led to an orientation based on purely mathematical structures—objects created within mathematics—and to the observation that geometrical intuition may lead to mistakes. The consequence, he wrote, was

> a radical treatment, the removal of sense and meaning from mathematics. What remains are mere objects of thought, with which one operates according to given rules, but which have no meaning in themselves.   [Bieberbach, AS 2b, 16]

Bieberbach conceded the value of formalist methodology, but called for a "meaningful" mathematics in which formalism should be embedded. These views place Bieberbach within a tradition that is still alive. His proposed remedy, however, was a regression to the mathematics of the nineteenth century, to the use of geometrical *Anschauung* not only as a concealed heuristic tool but as an acceptable form of presentation and argument.

The "loss of meaning" deplored by Bieberbach (and others) during the early twenties was very much a result of a gradual but basic change in the forms of scientific communication, namely in the language of mathematics. With modern axiomatics and logical symbolism, the formal justification of proofs and theorems had become the primary purpose of the language of mathematics. Imprecise elements—like imagery or motivational arguments that communicate some "meaning" other than scientific validity—were increasingly excluded. The official "modern" communication of a piece of mathematics presents a statement or a series of statements and, by the mode and means of expression, its rigorous justification, nothing more. Bieberbach observed—and attacked—what he saw to be the resulting deprivation of the content of communication, as given in natural—in contrast to formal—languages. Mathematics had indeed sacrificed "meaning" in order to attain rigor. Part of this development consisted of stripping the objects of mathematics to what Bieberbach in his lecture called

> pale skeletons in the sand of the desert, of which nobody knows what they mean or what purpose they serve.   [Bieberbach, AS 2b, 13]

The definition of these objects had been largely reduced to their mere formal elements. Furthermore, mathematics generated such objects itself, while remaining officially silent about the context of their genesis.

Bieberbach's complaints were not entirely rational; certainly his motives were not. Thus there is no point in debating the merits of formalism as related to rigor and creative freedom. The question is: which experiences led Bieberbach to his position, and how was this position connected to politics? The "modern" form of communication in mathematics (as described above) is an expression of the modern social system of mathematics. The form of communication determines a sharp boundary between the system and the outside, and it also tends to sharpen internal boundaries between specialties. This clearly defined external boundary contributes greatly to the professional autonomy of mathematics. No layman, e.g., in a ministry of education or research, can evaluate what mathematicians do or should do. Such professional autonomy is purchased with a loss of integration into society and with a differentiation of the social system. Socially separate fields for the pedagogy, philosophy, and application of mathematics become necessary.[6] Such structures and boundaries of social systems are not immediately visible, but they are strongly experienced in the forms and content of communication. Thus Bieberbach's attack on formalism was implicitly an attack on the structures of the social system of mathematics.

Bieberbach's arguments and appeals (in this lecture and elsewhere) were, like those of Boutroux and Klein, directed at a social integration of mathematics. Mathematical research, education, applications, and the closely related fields should not be permitted to become separated, and mathematics should make itself felt in the life of the nation. A philosophy like Brouwer's, in which the root of mathematics was to be found in a simple and basic human activity—namely to do one thing after the other—and which attempted to reconstruct the field so that any part of mathematics may be constructed from that root, could easily become a symbol for social integration. Brouwer's foundational papers, however, were rather formalistic pieces of mathematics. But this would not reduce their symbolic meaning. Bieberbach cared little for the mathematical technicalities of Brouwer's constructions; instead he used them as symbols that show the relation of mathematics to human life.

Hilbert's formalism, in contrast, appeared to be an attempt to build a wall of logical consistency around a paradise of free and esoteric mathematics, thus isolating it from matters of education, application, *Volk*, or nation. In this way the two "ideals" of mathematics are potentially linked to political values: social integration versus social differentiation, bonds to society versus professional autonomy, anticipation of an open future in which no problem is unsolvable versus the acceptance of restrictions in order to remain on the safe ground of reality. The conflict between these ideals and their relation to social and political debate was not unique to Germany; but after the humiliation and disillusionment of the German defeat and under the subsequent chaotic political and economic conditions of the early Weimar Republic, they were to have a profound emotional impact.

Before turning to more concretely political issues, let me note one of the ironies of history. In 1914 Bieberbach had ascribed intuitionist backwardness to the French; during the war the attack was returned. French authors wrote that the Germans were preoccupied with an excess of the *esprit de géométrie*, visible in their rigorous algebraic constructions based on arbitrarily chosen axioms. The Germans, they added, would look upon common sense with scorn, whereas the French realized that in reality the true basis of mathematics was the *esprit de finesse*. Pierre Boutroux's father, Emile, had been one of the voices in this chauvinist concert (cf. [Kleinert, 1979]).

Via Pierre Boutroux the basic dichotomy made its way to Bieberbach who, in his writings of the Nazi period, described French mathematicians as those "jugglers" with empty concepts, lacking in common sense. Remarkably, it appears always to be common sense—*Anschauung*—or a sense for reality that characterizes the right mathematics as part of the right nation.

Bieberbach's lecture of 1926 did not yet reveal nationalist or racist undertones; its political thrust remained with mathematics. That this lecture was given before the Association for the Advancement of Education in Mathematics and Natural Sciences—the so-called *Förderverein*—made it a matter of professional politics, since the *Förderverein* was fairly influential in educational politics. In the German system of higher education the main task of university mathematicians was to train secondary school teachers; thus the social status and legitimacy of mathematics rested strongly on the role of mathematics in secondary education. Therefore teachers, the *Förderverein*, and school mathematics traditionally played a major role in the professional politics of science.

By 1926 the times of external crises were over, and the political and economic situation had stabilized. In mathematics the foundational "crisis" had begun to recede as well. Although there remained a bitter animosity between Hilbert and Brouwer, the majority of mathematicians considered foundations to be one of many specialties of mathematics [Mehrtens, 1964, 261–262]. Working mathematicians were able to accept a degree of uncertainty in the logical foundations of their field, and the modern "style" prevailed in creative research. In contrast, Bieberbach appeared as a relic, stubbornly fighting for a cause that was no longer viable.

MATHEMATISCHE ANNALEN, BOLOGNA CONGRESS, ZENTRALBLATT

Revolutions and counter-revolutions are part of the struggle for political power and social control, within science as well as outside. Not only did Bieberbach attack Hilbert as the representative of the modern style of mathematics, he also joined forces with Brouwer in a futile attempt to reduce the influence of Hilbert and Göttingen mathematicians. Göttingen was the center of mathematics in Germany, leaving Berlin a distant second. Göttingen's preeminence had been established by Klein and Hilbert before World War I, and it was reinforced in the 1920s by Richard Courant's organizational activities, especially by his close cooperation with Ferdinand Springer, whose publishing house was then attaining the dominant position in publishing works in mathematics. In 1917 Springer had founded the *Mathematische Zeitschrift* and taken over the leading journal, *Mathematische Annalen*. At the same time Courant and Springer had started the "yellow series," the most important series of mathematical books of the period. Moreover, with funds provided by the International Education Board (Rockefeller Foundation), Courant had managed to build a new mathematical institute in Göttingen. By the second half of the 1920s there was a closely knit mathematical network—built on institutional, personal, and scholarly relations—the center of which was the Göttingen institute [Reid, 1970; 1976].

Consequently, Göttingen became the stronghold of mathematical modernism as well as the center of social power within the discipline. In Bieberbach's activities the struggle against the modern "ideal" of mathematics was always closely related to a struggle for power within the discipline. Bieberbach was not, however, a determined fighter; indeed he appears to have been a weak personality who sought an external source of strength to take a clear stand. In the conflicts with Göttingen (described below) Bieberbach chose to stand in the shadow of Brouwer. The latter had been involved in rejecting French attacks against German science during World War I; after the war Brouwer was active in fighting the international boycott against the Germans in science.

Brouwer cooperated with Karl Kerkhof, head of a "center for scientific information" in Berlin, who was prominent in German anti-boycott activities. On May 11, 1925 Brouwer [AS 1a] wrote to Kerkhof that on the question of what should be published in the "Einstein-Blumenthal matter," Kerkhof should turn to Bieberbach whose judgment he (Brouwer) would "accept offhand." Obviously Brouwer and Bieberbach were quite intimate on political matters, since the case in question was a delicate problem. The "Einstein-Blumenthal matter" referred to the problem of French contributions to a volume of the *Annalen* honoring Bernhard Riemann. Both Bieberbach and Brouwer, who were on the editorial board of the journal, opposed the decision of the managing editor, Otto

Blumenthal, to invite the participation of French authors without regard to their political positions. Blumenthal had, in this question, relied on Einstein's good relations with French scientists, and this led to the involvement of Paul Painlevé. Brouwer protested sharply, quoting at length Painlevé's anti-German statements from the war years. Bieberbach then proposed a compromise, namely that Einstein should approach selected French mathematicians, Painlevé excluded [Einstein AS 1a]. The information was obviously passed on (as the "Einstein-Blumenthal matter") to Kerkhof, who eagerly collected and published material on anti-German politics in science.

Bieberbach's compromise had solved the problem for the *Annalen*, but there remained a struggle for influence within the editorial board. Hilbert, Blumenthal, and Carathéodory were the principal editors; the others, including Einstein, Brouwer, and Bieberbach, "cooperated." In February 1925 Hilbert wrote to Einstein:

> While I presently consider the matter of admission to the Riemann-volume as definitely settled, I received a rather superfluous letter from Bieberbach on the question.   [Einstein, AS 1b]

In this letter Bieberbach demanded a unanimous vote from the editors—not just the three main ones—on the decision. In all likelihood Hilbert suspected Brouwer of being the driving force behind this request. In fact, the conflict between Brouwer and Hilbert was growing to extremes, and although Bieberbach apparently sided with Brouwer, he sought to resolve matters by compromise.

When in late 1925 Bieberbach requested 300 free copies of the second edition of his book [1923] from Springer [Bieberbach, AS 7], Courant—suspecting that someone probably had incited Bieberbach to take this "grotesque" action—proposed a moderate response to the request, explaining that

> Bieberbach has a certain influence through his position in the mathematicians' association and as a Berlin bigwig, and, even more through his fresh, impulsive manner.   [Courant, AS 1a]

Later, when Bieberbach asked for access to the proofsheets of the *Annalen*, Courant commented,

> Possibly Hilbert as the principal editor does not wish that Bieberbach should attain a special position through which he could obtain new privileges. In

general Hilbert finds himself of late occasionally in a defensive position with regard to Herrn B.   [Courant, AS 1b]

The climax in the struggle around the *Annalen* was precipitated by another event. Italian mathematicians, under the chairmanship of Salvatore Pincherle, were preparing for an International Congress of Mathematicians with German participation, to take place in September 1928.[7] As early as 1925, when the first plans for the congress were being made, German mathematicians became concerned about the involvement of the *Union Mathématique International*. The Union was closely related to the *Conseil International des Recherches*, which had been mainly responsible for the international boycott of German science ([Schröder-Gudehus, 1966], [Forman, 1973], and [Cock, 1983]). When invitations to the congress were released, it was not entirely clear that the organization of the congress was completely separated from the Union. Furthermore, the invitations contained an announcement of an excursion to the "liberated area" of Southern Tirol. This was a political affront to many Germans, and Brouwer, after failing in his attempts to renegotiate arrangements in Italy, prepared an appeal for a counterboycott. Bieberbach, as secretary of the DMV, wanted to maintain a neutral stance, since the DMV had decided not to send an official delegation but rather to leave decisions about participation up to its individual members. Nevertheless, in a letter that was very likely never intended to be widely publicized, but was circulated in the German universities, Bieberbach wrote that he would "deeply regret large scale attendance of the German mathematicians." Citing the planned excursion to Southern Tyrol as a particular example of "tactlessness," he accused the Italians of "veiling" the fact that the congress was organized by the Union. Hilbert responded in a circular to this "denunciation of the congress" by arguing that Bieberbach's position had no factual basis and the invitation of German mathematicians to Bologna should be accepted. A handwritten note, filed with the circular, reveals Hilbert's suspicions:

In Germany there has arisen a [form of] political blackmailing of the worst kind: You are not a German, unworthy of German birth, if you do not talk and act as I prescribe. It is very easy to get rid of these blackmailers. You have only to ask them how long they were in German trenches [in World War I]. Unfortunately, German mathematicians have fallen victim to this blackmailing, Bieberbach for example. Brouwer has managed to take advantage of the situation of the Germans without having been active in the German trenches himself. He has cultivated this instigation of discord among the Germans, all the more in order to set himself up as the master of German mathematics. With complete success. He will not succeed a second time. [Hilbert, AS 1]

Brouwer, whom Hilbert saw as the blackmailer, distributed a printed circular of his own calling for a boycott of the congress. Many mathematicians were involved in discussions and negotiations, among them Harald Bohr from Denmark and Godfrey H. Hardy from England. From the correspondence it is obvious that while the foreigners tried to mediate the situation, the parties in Germany were badly split: Brouwer (although Dutch) stood against the congress; he was joined by the Berlin mathematicians, namely Bieberbach, Richard von Mises, and Erhard Schmidt; the Göttingen group, in contrast, favored participation. Obviously, local competition, political position, and personal conflicts were all at stake here. In the end Hilbert led a large German delegation to the congress, and in his opening address he made a strong plea for international cooperation, which was greeted with tumultuous applause.

At this time Hilbert was seriously ill and, as Carathéodory informed Einstein, feared that after his death Brouwer might be a danger to the further existence of the *Annalen*.[8] Soon after the Bologna Congress Hilbert asked the other editors to agree upon Brouwer's elimination from the editorial board. Ten days later he wrote a letter informing Brouwer that the principal editors had agreed to terminate his association with the *Annalen*. It was reported that Brouwer never even opened this letter. Another period of hectic activities followed, a tragicomical story, that Einstein documented in a separate file entitled *Der Frosch-Mäusekrieg*.[9] In the end a new contract with the publisher made Hilbert the editor and named Hecke and Blumenthal as collaborators, thus eliminating both Brouwer and Bieberbach from the board. A year later Bieberbach appeared on the editorial board of the *Mathematische Zeitschrift*, which may well have been a move to consolidate the relations between Springer and the Berlin mathematicians.

In the course of events Brouwer and Bieberbach paid a visit to Springer, and in a note on the conversation [Einstein, AS 1b] Springer wrote that his visitors had informed him that

> attacks could be expected on the publisher, who could develop the reputation among German mathematicians as one who lacked a proper regard for his country.

Furthermore, Brouwer threatened to start a competing journal with another publisher; Bieberbach added that he would leave the *Annalen* if Brouwer was removed.

In 1930 Brouwer did, in fact, start a new journal in the Netherlands, the *Compositio Mathematica*, in which Bieberbach was involved from the very beginning. But a new conflict involving the same persons and institutions soon arose. Bieberbach was chairman of the committee of the Prussian Academy of Sciences which was responsible for the *Jahrbuch für die Fortschritte der Mathematik*,[10] the leading abstracts journal of the day. Since World War I the *Jahrbuch* had suffered from a serious lag between the time of publication of articles and the appearance of the abstracts. In 1928 the volumes containing abstracts of publications for 1923 and 1924 were prepared. Bieberbach was actively engaged in the effort to overcome the shortage of editorial staff and the problem of timely and complete acquisition of review copies from the publishers. The managing editor, Georg Feigl, noted in December 1930 in his annual report that he had privately received news that Springer, in cooperation with a number of Göttingen mathematicians, was planning to begin publication of a new abstracts journal early in 1931. Springer also had offered to take over the *Revue Sémestrielle* from the Dutch Mathematical Society.

In response the editors of the *Jahrbuch* decided to engage more personnel and to discuss with Brouwer the possibility of cooperation between the *Jahrbuch* and the *Revue*. Through Brouwer's influence the two journals merged in 1932, but for financial and political reasons this cooperation ended by late 1934. Meanwhile the *Jahrbuch* editor received official notice of Springer's plans in November of 1930. Various discussions about possible cooperation or division of labor led nowhere. Even the DMV—with Bieberbach still as secretary—intervened in favor of a fusion. But in 1931

Springer started his *Zentralblatt*, with Otto Neugebauer as managing editor and an impressive array of editors and collaborators.

For Bieberbach another battle was lost, and the survival of the *Jahrbuch* was seriously in question, although it managed to survive through the following decade. Neugebauer left the *Zentralblatt* in 1938 when the name of a Jewish editor (Levi-Civita) was, without further notice, eliminated from the title page. When *Mathematical Reviews* first appeared in 1940 cooperation between *Jahrbuch* and *Zentralblatt* was enforced in Germany. The *Jahrbuch*, however, could not be revived after the War, whereas the *Zentralblatt*, despite American opposition, had a successful new start (cf. [Siegmund-Schultze, 1984a]).

By 1930, Brouwer had withdrawn from the foundational debate, and in 1934 he passed up the chance of a belated but doubtful triumph over his adversary Hilbert by declining the offer of a chair from Göttingen. The few documents relating to the offer reveal that at first Brouwer showed interest, but let the matter end, officially for financial reasons. Bieberbach was now isolated from the leading, Göttingen-centered, group in mathematics. At the reception celebrating Hilbert's seventieth birthday in 1932, it is said[11] that hardly anybody bothered to speak to him. He certainly had many colleagues who sympathized with his interests, especially in Berlin, but times were changing. On April 18, 1933 Bieberbach sent a picture postcard with birthday greetings to his Berlin colleague von Mises, who had been on his side during the Bologna affair. The postcard showed a small German town over which a stylized sun inscribed with a swastika had risen. Von Mises was of Jewish descent, and his response to Bieberbach's greetings is not known. But on May 1 he sent a postcard to his mother showing Hitler, Hindenburg, and a worker, all in heroic poses; "Day of Labor" was printed below. On the back von Mises wrote the names of those colleagues who had been dismissed in Berlin [von Mises AS 1].

DONNING THE BROWN SHIRT

Bieberbach's turn to Nazism came as a surprise to most of his students and colleagues. He had been seen as a "good republican"

[Gumbel, 1938, 255], "liberal and democratic" [Rado, AS 1], certainly "no anti-Semite in the usual sense" [Fraenkel, 1967, 152]. Hans Freudenthal [AS 1] recalls that Bieberbach was reputed to be the most leftist of the Berlin mathematics professors—one could even meet communists in his house. In his political activities and ideas on mathematics, psychology, and race, Bieberbach was generally considered to be "subjectively honest" [Weyl AS 1a]. After the War Courant wrote,

> As to individuals I do not think that anyone behaved as crazily as Bieberbach; I cannot help feeling that he is and always was just crazy but not really dangerous.   [Courant, AS 2]

But the evidence reveals that Bieberbach was a staunch nationalist and that he had very strong opinions on how mathematics should be done. Although it is possible his actions in the cases of the Bologna Congress and the *Annalen* were largely incited by Brouwer, Bieberbach surely experienced their outcomes as a personal defeat. Moreover, in his thinking and activities he did not separate general politics from disciplinary politics, nor did he make a distinction between social aims and values and those based on professional standards. At the very least he was prepared to build up even stronger political connections and, using political power, to take up the struggle for professional dominance. What remains to be explained is his willingness to join forces with the Nazi regime and to do so in such an uncompromising manner.

There were some obvious discrepancies in Bieberbach's behavior: the treatment of his Jewish friends and colleagues probably posed the most difficult problem, and here, like most Germans at that time, he was able to overlook what was going on around him. The suppression of left-wing opposition was another matter, and the bourgeois German academics preferred the brown "barbarians" to the danger of the "red chaos." Hoping that the authoritarian regime would restore Germany's pride and power, they accepted the barbarian outgrowths as a passing necessity of the Nazi "revolution." The politics of the Jewish question was different: it symbolized a volkish tradition of authoritarian power in which the "alien" had always been defined in an arbitrary way. Anyone who wanted to believe in and to find his place in this "New Germany" was forced to shut his eyes and to ignore the terror that the Nazi regime unleashed.

Bieberbach, it seems, did want to believe in the new order. One of the earliest clear signs of his turn to Nazism was, much to the mockery of his colleagues, his participation in a highly publicized march of storm troopers (SA) from Potsdam to Berlin. Bieberbach marched along with his sons, who were at this time 11, 15, 17, and 18 years of age. In November 1933, Bieberbach joined the SA, where he did not try to evade physical activities, indeed, he received the SA sports badge, which indicates the extent to which he participated in the Nazi "revolution." The good order it was supposed to infuse in German life also extended, for Bieberbach, to mathematics. Again in July 1933 he lectured on *Anschauung* and thinking in mathematics. To the topics of his 1926 lecture he now added that "German mathematics" would recover its proper style. He also alluded to the racial roots of this "style," relating it to Felix Klein's remark of 1893:

It would seem as if a strong naive space-intuition were an attribute pre-eminently of the Teutonic race, while the critical, purely logical sense is more developed in the Hebrew races.   [Klein, 1911, 46] (see also [Rowe 1984])

Later that year Klein, Bieberbach's hero, came under attack. In November 1933 the Nazi physicist Philip Lenard distributed a memorandum written by the philosopher Hugo Dingler to the Prussian Ministry of the Interior. The document was a grotesque racist attack on physics and mathematics in Germany, which were allegedly dominated by a Jewish conspiracy centered in Göttingen where Klein had been the ring leader [Dingler, AS 1]. A lengthy article by Eva Manger [1934a], who apparently was a student of Bieberbach, demonstrated Klein's purely Aryan descent to the DMV membership. The attacks on the "Jew" Klein, however, were closely connected to a general attack on mathematics. Lenard, who was supporting Dingler, had tried to make himself the prime referee for all university appointments in mathematics and physics [Kleinert, 1980, 35]. At the same time Lenard promoted the idea that the proper, "German," physics was not in need of any more than elementary mathematics [Lindner, 1980, 91]. This provided Bieberbach with a motive for the construction of a "Deutsche Mathematik": the defense against Nazi attacks on mathematics. His address to the *Förderverein* in April 1934 (see below) served, in his own words, to show that mathematics did not need further legitimation because it was a matter of national heritage [Bieberbach, 1934c, 243]. His later correspondence with Dingler also shows that he felt very strongly about the attack on Klein [Dingler, AS 2].

The *Förderverein* lecture was the subject of the newspaper report quoted in the introduction to this paper. Reactions to this report led to Bieberbach's conflict with the DMV—resulting in his eventual exclusion from the society—and to the foundation of his journal, *Deutsche Mathematik*. By this time Bieberbach was well-established as a National Socialist activist. In November 1933 Bieberbach joined the National Socialist Lecturers Association, and a little later became the representative of this organization at the University of Berlin. In January 1934 he and Eduard Wildhagen applied for a grant to do secret military research. Through Wildhagen he maintained relations with the then still powerful SA leaders and the Nazi physicist Johannes Stark, who had become president of the German Research Association (DFG) [Heiber, 1966, 791ff]. At this time Bieberbach was appointed deputy to the Nazi rector of Berlin University, the "race scientist" Eugen Fischer, who later proposed Bieberbach as his successor. Although he was not elected to this position, Bieberbach became the permanent dean of his department [Grau, 1979, 28]. Through these positions Bieberbach had close contact with Theodor Vahlen, the first mathematician to become a member and functionary of the Nazi party during the mid-twenties and, beginning in 1933, a leading figure in the central university administration in the Nazi ministry [Siegmund-Schultze, 1984b].

In 1933–34 Bieberbach left the editorial board of *Mathematische Zeitschrift*, probably because of the "Jewish" publisher, Springer.[12] He also tangled with Brouwer about the participation of Jewish editors in *Compositio Mathematica*, and by 1936 all "real" Germans had left the editorial board of the Dutch journal; these included several emigrant Germans who were, by Nazi standards, Jews.[13] The managing editor of the *Jahrbuch* was ordered by Bieberbach to alter its policy regarding Jewish contributions. Bieberbach wrote that he had so far not objected to occasional Jewish participation, but the latest issue showed a "grotesque" collection of "Jews and their notorious friends" [Freudenthal, AS 2]. If Bieberbach was not a convinced anti-Semite he certainly fulfilled his functions as a willing servant of the anti-Semitic regime.

Bieberbach's conflict with the DMV started when, against the wishes of his co-editors, he published an "Open Letter to Harald Bohr" [Bieberbach, 1934a] in the *Jahresbericht*. This was in response to an article [Bohr, 1934] published in a Danish newspaper about Bieberbach's *Förderverein* lecture. Bohr had sent this article to several German mathematicians. Bieberbach defended his position by claiming that he was merely calling on the inner strength of a *Volk* to become conscious of their proper style of thinking, adding that differences between mathematical styles were obvious. He attacked Bohr as "a pest to all international cooperation," whose writings revealed his "hatred of the new Germany."

With this "Open Letter" Bieberbach, as the secretary of the DMV, created the impression that his views represented those of the society. Since the DMV had a large foreign membership this tirade endangered its international position. Further, Bieberbach—with his intimate connections to Vahlen and other Nazi officials—was an immediate channel for Nazi ideology and power to intrude into the affairs of the society. The matter was put on the agenda of the DMV's annual business meeting, held in Bad Pyrmont in September 1934. At the outset of the meeting the chairman was brave enough to invoke the statutes in order to exclude all non-members, namely the student-followers of Bieberbach who had appeared in SA uniform. When the Bohr case came up, however, neither the chairman nor the DMV membership were sufficiently courageous or resourceful to handle the situation that arose when Bieberbach's Nazi followers gave the matter an unexpected turn. Rather than posing the question of the acceptability of Bieberbach's behavior—as an officer of the society—they raised the point that

Bohr's article was an attack on National Socialist Germany which Bieberbach had to answer. Thus a resolution was passed in which the DMV merely "regretted" Bieberbach's behavior and his letter, but "condemned" Bohr "in so far as one can see in his article an attack on the new German state."[14]

Bieberbach's attempt to have himself installed as "Führer" of the DMV, however, was voted down by a large majority. Instead, an ingenious amendment to the statutes, proposed by Erich Hecke and passed by the assembly, gave the impression that the DMV had shown its loyalty to the Nazi system by adopting a moderate form of the leadership-principle: the new chairman would have the power to appoint and dismiss all other officers. The primary motive for this amendment, however, was not an adherence to the leadership-principle, but rather the possibility of getting rid of Bieberbach, who could not be dismissed under the old statutes. With this result in hand Hecke went to visit Harald Bohr and urged him not to resign his DMV-membership. Hecke promised that the resolution concerning Bohr would not be publicized [Bohr, AS 1].

The preparation of the minutes of the meeting, however, fell to the secretary, Bieberbach, who published—contrary to Hecke's promise—a very detailed version of the minutes that included the resolution on the Bohr-Bieberbach matter. Furthermore, before the amendments could go into effect they had to be filed, and Bieberbach hesitated to do this. The other officers took the opportunity to accuse him of behavior unworthy of a colleague and publicly requested his resignation. In the course of this conflict it emerged that the amendments were incorrect in certain details and, therefore, could not be filed. Thus Bieberbach could argue that he had hesitated because he had anticipated such problems. At that point he attempted to turn the attack around, by using his connections and the ministry, to put pressure on the DMV. However, neither the ministry nor his colleague, Hamel, gave Bieberbach the backing he wanted. In the end both Bieberbach and the chairman of the DMV had to resign their offices. In an informal agreement the DMV declared its loyalty to the ministry, but this did not require the DMV to include the leadership-principle in its statutes.

Bieberbach realized immediately after the meeting in Pyrmont that he had no chance of dominating the DMV. Only a month later he applied for a grant to found a new journal,[15] which

> shall pave the way for the organization of a German mathematical research-community. ... It fills a long felt need by bringing together Volk and science, volkish origin and scientific achievement in the common work of university teachers and students.

Johannes Stark, then still president of the DFG, the funding organization for science, provided ample support. Although the publishing house Hirzel initially estimated the number of potential subscribers at 500, the journal began publication with 6500 copies and six issues a year. By publication of the fourth issue the number was reduced to 2000. The price, according to Hirzel, was half of what a standard publisher's calculation would have yielded. Yet by the first issue of 1938, only 700 copies appeared, of which 167 remained unsold three years later.

State or party power had not helped to bring the journal into a dominant position. On the contrary, after Stark was forced out of office (due to political intrigues) in November 1936 ([Heiber, 1966, 843], [Beyerchen 1977, 119ff]), his follower Mentzel, who also represented the ministry, very soon asked for a reduction of the funds provided by the DFG for the journal. Bieberbach defended what he had, especially the remuneration that went to editors and authors, and denounced the other journals:

> One (*Mathematische Annalen*) is edited by a Jew. In another (*Mathematische Zeitschrift*) there appear papers dedicated to female Jewish communists. In a third (*Crelle's Journal*) papers by emigrants are printed. A fourth (*Quellen und Studien*) is led by a Jew and an emigrant half-breed.

(The persons alluded to are, in order, Otto Blumenthal, Emmy Noether, Richard von Mises, Otto Toeplitz, and Otto Neugebauer.) Bieberbach's argument, that an upright German comrade should not publish in such journals, did not seem to help much. Funds were considerably reduced and a note in Mentzel's files reads as follows:

> The editors renounce a remuneration. The journal shall in accord with its aims be fully supported by the political will and the spirit of sacrifice of those concerned.

Obviously the representatives of the state did not expect much from Bieberbach's *Deutsche Mathematik*. The number of copies printed was afterward reduced to 700, and the journal continued publication until 1944 with a constant subsidy from the DFG. The plan for a new association of mathematicians was never realized. Responsibility for the journal was carried by a group of Nazi oriented mathematicians, mainly in Berlin and centered around Bieberbach, who were isolated from the larger mathematical community. The nominal editor was Vahlen, while Bieberbach acted as managing editor (*Schriftleiter*). But in reality it was Bieberbach, the founder of the journal, who dealt with its authors, the publisher, and the DFG. Apparently Vahlen's name was put at the top of the title page to ensure maximum official recognition for the journal, for Vahlen was, at this time, head of the department for science in the newly founded Reich Ministry of Education and Science. However, he left this office on January 1, 1937, probably in the course of the intrigues and infighting that led to Stark's fall [Siegmund-Schultze, 1984b, 26].

Bieberbach, with his *Deutsche Mathematik*, suffered the fate of many a "revolutionary" of the Nazi era. He had been instrumental in driving the DMV into the arms of the National Socialist state. The mathematicians feared the radical Bieberbach and thus were forced to declare their loyalty to the apparently less radical agents of the regime in the ministry of culture. In this way Bieberbach had played his small role in consolidating the regime, but the revolution in mathematics he had hoped for did not take place. The Nazi regime needed the revolutionaries and radicals to establish itself firmly, but revolutionary aims were soon set aside when the regime tried to consolidate and expand its power. The preparation for war required loyal and able specialists, not ideologues and political radicals. The DMV, on the other hand, had helped to defuse the radical movement Bieberbach had hoped to set in motion, forcing him and his group into a residual position without political influence. As Timothy Mason [1978, 106] has put it,

> Conservative forces preserved social and state organization for National Socialism by saving them from National Socialism.

TYPES AND RACES

In an apologia written in 1949 Bieberbach related his interest in styles of mathematics to the work of Poincaré and Boutroux, adding

> When in 1933, in the course of political events in my fatherland, I came to
> know the older typologies of race-science and psychology, namely the works
> of Erich Jaensch, I believed I had found the appropriate means to describe
> and perhaps even explain my observations. At the same time I hoped these
> studies would kindle the interest of those in power for scientific matters so
> that I could find some protection not only for myself.   [Hopf, AS 1]

These two elements, race-typology and the interests of "those in power," first appeared in Bieberbach's publications in 1934. The most important of these is the *Förderverein* lecture, published in two versions [1934b and c]. Also published were a lecture to the Berlin Academy [1934d] (which in essence repeats the earlier lecture); a brief paper [1935] on the history of differential calculus; and a lecture, given in 1939 at Heidelberg, which presents the most detailed description of his typology [1940]. Of the briefer statements, his preface to the first issue of *Deutsche Mathematik* [1936a] and a lecture on mathematics education in the universities [1937a] should be mentioned. In 1938 Bieberbach published two historical books, one on Gauss [1938a], which remains very close to the historical sources and makes very little use of ideology in its language, and another on *Galileo and the Inquisition* [1938b], a more popular little book with some anticlerical undertones and a message summarized by its last sentence:

> Thus for us as well, living in the present age, the freedom of science is an
> unalienable treasure and a compelling duty to the people, whose welfare is
> the beginning and end of all our action and thought. [Bieberbach, 1938b, 140]

"Free science in the service of the people" was a slogan in frequent use by 1937–38 [Mehrtens, 1980, 49]. Usually it meant nothing more than an ideologically unhampered science in the service of preparation for war. After the ideological attacks of *Deutsche Wissenschaft* during the first years, natural scientists and mathematicians used this slogan to legitimate their disciplines by citing their applicability in military and industrial technology. Bieberbach, in contrast, attempted to justify pure mathematics directly by racial and volkish integration, making remarkably little use of the application argument. In the lecture he gave on mathematics education at a National Socialist students' camp in the

Philip-Lenard-Institute at Heidelberg he stressed that to do
mathematics for its own sake, as in a field like number theory, was
a German characteristic and served the honor of the German
people. Bieberbach's efforts to counteract the influence of the
Deutsche physicists become clear in his closing comment:

> My remarks, I hope, will persuade Herr Lenard, the founder of this house, to
> adopt a more favorable judgement of the aims and worthiness of our
> German mathematical science. [Bieberbach, 1937, 16]

Here as in other publications Bieberbach stressed the importance
of the education of *Anschauung*, just as he had done prior to 1933.
Now, however, his basic argument was that it was the power of the
true German which had to be developed. This he repeated time and
again, especially in more defensive pieces like the Bohr letter. A
good example is the following passage from his statement of
editorial policy in *Deutsche Mathematik*:

> We serve and foster the German character in mathematics. We are not alone
> in the world: other peoples have the same right to the development of their
> own characteristics in their mathematical work. Manifold contacts exist
> between the mathematical work of different peoples. Our journal has an open
> mind to the resulting information and inspiration, but our pervasive point of
> view is that of the mathematical achievements of our people. This is the aim
> of our work, keeping in mind that mathematical creativity develops the more
> powerful...the deeper it roots in the heritage of a people. [Bieberbach,
> 1936a]

What Bieberbach meant by the type of mathematics that is
characteristically "German" is what he had already described in
1926, i.e., the "sound," "fruitful," and "organic" scientific ideal of
Felix Klein. His favorite example was still Gauss' "organic and
concrete" introduction of the complex numbers, as contrasted with
the "symbolism" of Cauchy and Coursat [Bieberbach, 1934c, 237].
Another prime target, Edmund Landau's definitions of π, sine, and
cosine by means of power series expansions, was branded "un-German." Landau's definitions lacked "sense and meaning," and his
"style" was contrasted with the presentation of the same concepts
by the "German" Erhard Schmidt, which was "organic," "concrete," and "anschaulich" [Bieberbach, 1934c, 236f].

In the introductory remarks to his 1934 lecture before the *Förderverein*, Bieberbach legitimized the Nazi-organized boycott of Landau's classes:

> A few months ago differences with the Göttingen student body ended the teaching activities of Herr Landau. ... This should be seen as a prime example of the fact that representatives of overly different races do not mix as students and teacher. ... The instincts of the Göttingen students felt [sic] that Landau was a type who handled things in an un-German manner.
> [Bieberbach, 1934c, 236]

In an extant manuscript of this lecture [Bieberbach, AS 3a, 14] the actions of the Göttingen students are characterized as "clever," "justified," "deserving acknowledgment," and "manly." This passage did not appear in print; instead, a few lines below, Bieberbach inserted in the published version the remark that the argument concerned the "mode and style," not the "unquestioned merits" of Landau in finding "new scientific facts" [Bieberbach, 1934c, 237]. The differences between manuscript and publication are not extensive, but there is a change of tone, due perhaps to the critical reactions following the newspaper report of the lecture.

The boycott of his classes forced Landau, who formally was not subject to the first wave of dismissals, to resign his position [Kluge, 1983]. Bieberbach's speech not only justified this boycott but provided a rationale for the dismissal of Jewish scientists in general. To gain the interest and the support of "those in power" he had to take part in the justification of their actions. There is no evidence that Bieberbach was actively involved in the very early and radical purges of Göttingen's institute of mathematics and physics, but his long-standing opposition to Göttingen may well have been a reason for his attack on Landau. Further, it is reported [Ostrowski, AS 1] that Landau had openly expressed his contempt for Bieberbach. Letters from Landau to Bieberbach [Bieberbach, AS 3b] reveal the arrogant manner in which Landau pointed out mistakes and possible improvements in Bieberbach's work. The "inhuman" qualities Bieberbach ascribed to Landau's mathematical style were, perhaps, in return for the humiliations which Bieberbach could not counter by better mathematical results.

Hilbert, although too old to play an influential role at this time was too monumental a figure to attack. Thus Bieberbach had the difficult task of placing the formalist Hilbert in his typology:

> The struggle over the foundations of mathematics is dependent on races or, expressed differently, one's position in this struggle corresponds to certain types of intellectual creativity. As such the J-type will tend towards intuitionism or towards the ways of Klein, while formalism appears to belong to the S-type. This seems to be contradicted by the East-Prussian heritage of the founder of formalism. Indeed Hilbert cannot possibly be taken as an S-type regarding his other accomplishments. In the psychology of types, however, a form of J-type is known which tends to be open to influences of the S-type. [Bieberbach, 1934d, 10]

Further explanation is not given, and the reader had to place his faith in Bieberbach's ability to master the finer problems of typology. Five years later Bieberbach repeated the typological interpretation of the foundational debate, but this time he found another solution to the Hilbert problem:

> The difference is quite compatible with the fact that both Hilbert and Brouwer should be classified under the psychological type J_3/J_2. The fact that two men approach their science with an ideal norm does not necessarily imply that it has to be same norm in both cases. [Bieberbach, 1940, 27]

In the face of such "juggling" one might well imagine that Bieberbach may have had some difficulty believing what he said and wrote. Still, all the evidence suggests that he was successful in doing so!

The typology Bieberbach drew upon was borrowed from the psychologist Erich Jaensch,[16] who had written on the structure of human cognition, especially the perception of space, and had attempted to develop a scientifically based typology. In his lecture of 1934 Bieberbach quoted Jaensch's book [1931] on cognition explicitly. A typological study of mathematical thinking had been prepared by a student of Jaensch in 1930–31, but only published in 1939 [Jaensch and Althoff, 1939]. The published version quoted Bieberbach's typology and stressed the agreement of the "independently" achieved results [Jaensch and Althoff, 1939, 73]. Jaensch had postulated two basic psychological types, the "S-type" with unstable psychic functions, internally generated synaesthetic (hence the "S") perceptions, and a tendency towards disintegration; in contrast was the "J-type" with stable psychic functions, in whom perceptual imagery and conceptual thinking were strongly integrated (the "J," earlier "I," is related to the term "integration-type").

What made Jaensch's ideas attractive to Bieberbach was not only the basic polarity of types and the discussion of perception, imagery, and thinking, but also the values and associations that accompanied the typology. Like Bieberbach, Jaensch insisted that he was working "scientifically," while at the same time he was drawing general political consequences from his theories. In fact he connected his ideas with a general critique of modern culture, which he found disintegrating as it led to a growing separation of intellect and soul. Jaensch greeted the Nazi movement as a cultural renaissance that would lead to the dominance of a sound type in the sense of his theory. His path to National Socialism was not much different from that of Bieberbach [Geuter, 1985].

Bieberbach adopted Jaensch's two basic types and the subdivision of the J-type into J_1, J_2, and J_3. In general the J-type was, in Bieberbach's words,

> open to reality, with all senses and psychic functions, so that for him *Anschauung* and thinking merge into a harmonic unity.   [Bieberbach, 1934d, 5]

The differentiation of types appeared later in Bieberbach's Heidelberg lecture:

> The J_1-type does not turn the world into a problem, rather the problem comes to him out of the world. ... He is attracted by the colorful richness of reality; he is interested in coherence, in the grand scale of events; while thinking he has to see or feel the relation to reality.   [Bieberbach, 1940, 15]

For Bieberbach, Felix Klein was the paradigmatic J_1-type, while Gauss represented the J_2-type, which

> does not so much long for a wealth of knowledge but rather for its meaning and range. He approaches reality with fixed values and ideals. He tries to form cognitions into a world-view. His aim of work is a perfect harmonious construction. He loves truth for its beauty.   [Bieberbach, 1940, 15]

The third type, J_3, is illustrated by Weierstrass:

> It is the type for whom knowledge must have command over the world. ...
> The scientist and his subject matter are standing face to face like fighters

struggling for power. Cognition is a struggle with reality. In pure mathematics these are the critics, the systematists, who carve out clear rules for the control of the subject, who clarify the basic concepts and deprive them of their mysteries, who form the accumulated results into a system.   [Bieberbach, 1940, 16]

In contrast, there is the S-type:

None of these types runs the risk of accepting no criterion other than the inner coherence of the edifice of his thought or losing sight of the natural place of things in science. This is in fact the case with the fourth type, which Jaensch called the S-type, the *Strahltypus*, who beams his autistic thought into reality. At best he tries to recover his ideas within reality, but not as a confirmation of his thinking, rather as an *epitheton ornans* of reality. Among the great German mathematicians—I emphasize the word 'great'—no case of this intellectualist type can be found. Among Germans, however, juveniles and also mathematicians frequently remind one of this type. Namely, among the strangers who took up certain studies of Hilbert, there are some who belong to this intellectualist type. [Bieberbach, 1940, 16f]

("Strangers," in this quote refers not only to foreigners but also—and in a stronger sense—to the racial "strangers.")

In his publications Bieberbach correlates these psychological types with racial types. The "Nordic" race was most clearly represented by J_3, while "Eastern" and "Oriental" races tended to the S-type. Weierstrass and Gauss are labelled as "Nordic-Falian", while the "Nordic" Klein revealed a "Dinarian" element. Euler was "Eastern-Dinaric" and Jacobi "Oriental" [Bieberbach, 1934d].

The first reports of Bieberbach's remarks were received abroad "with varying degrees of sorrow, derision and contempt," as Oswald Veblen wrote to Bieberbach [Veblen, AS 1]. The principal reasons for this outcry were Bieberbach's justification of Landau's dismissal and the distinction he had made between "German" and "Jewish" mathematics. To the latter reproach Bieberbach reacted promptly through his spokesman and student, Eva Manger, who sent a letter to the editor of the newspaper that had reported on the *Förderverein* lecture:

Bieberbach has in no way committed himself to the unfruitful distinction between 'Aryan' and 'Jewish' mathematicians. The word 'Aryan' was not used at all in his lecture.   [Manger, 1934b]

She repeated Bieberbach's (and Jaensch's) contorted argument that

the rejection of alien characteristics was not meant as a denouncement of the personality and achievement of those who could not be
counted as "German."

In fact, neither the publication nor the manuscript contain the
word "Aryan"; the words "Jew" or "Jewish" appear only twice in
the published text in quotations from other authors, while in a
footnote Bieberbach added that a "Jewish mathematician like
Minkowski or a French mathematician like Dirichlet" might have
clear traits of the J-type [Bieberbach, 1940, 241]. A sentence with
the phrase "Jewish thought" is crossed out in the manuscript itself.
Even in his lecture at the National Socialist stronghold Heidelberg,
Bieberbach warned against assuming that "every German is a
J-type and every Jew an S-type" [Bieberbach, 1940, 18]. Similar
care was taken with French mathematicians who were not, as a
group, labelled typologically.

In his later apologia Bieberbach wrote:

> Only much too late did I realize that in these investigations I was playing
> with fire and that they brought me into the neighborhood of bad company,
> where the good I perceived among other things was merely a mask for
> criminal and demagogical aims.   [Hopf, AS 1]

It must have taken considerable self-deception not to see the "bad
company" for what it was. For example, as editor of *Deutsche
Mathematik*, Bieberbach sanctioned an article by Erhard Tornier,
which appeared only a few pages after the carefully worded introduction to the first issue. Tornier took up the juggler metaphor
Bieberbach earlier had borrowed from Boutroux and referred to

> Jewish-liberalistic smoke-screening, springing from the intellect of rootless
> artists who, for their regular public, conjure up mathematical creativity with
> definitions alien to the object... [Tornier, 1936]

Perhaps Bieberbach felt it was necessary, for the sake of "the
good" he saw in the movement, to allow a certain amount of such
verbal radicalism. His own "scholarly" treatment of the matter
made him immune to his responsibilities and prevented him from
seeing the consequences of his utterances. His typological "scholarship" shows a considerable lack of critical judgment and awareness
of human realities, and it embodies an extreme, if by no means
uncommon, lack of humane qualities in political judgment.

The emotional basis for Bieberbach's convictions is grounded in beliefs that may be traced to his early career. His fight against formalism and his experience of its dominance as a matter of social power centered in Göttingen explain, in part, his willingness to use political power in an attempt to recover what he felt was a sound order in the world of mathematics. Mathematics was his world and he suffered greatly from the apparent loss of meaning, coherence and harmony after so many rapid changes. This feeling of imbalance was present in many groups whose worlds seemed to fall apart during the political and social upheavals of the first decades of the century, and it became a strong impetus for the fascist movements of those times.

The longing for harmony and social identity was not untainted by self-interest. Indeed, in Bieberbach's case, it was connected with the desire to attain a safe and strong social status. He wanted to dominate German mathematics by forcing it to adopt his views on what it should be, and he wanted to ensure mathematics (and himself) a high status in the National Socialist world. But, in the end, he was deceived by that movement which was carried to power by anti-modern sentiments and regressive hopes for identity and harmony. Once in power, the National Socialists had to seek some compromise with the existing powers of the modern, industrial state; it used the most modern means of propaganda and terror to remain in power. Its aggressive imperialism unchained the destructive forces of modern technology and bureaucracy. National Socialism meant not only radical politicizing and reactionary political romanticizing, but unpolitical professional specializing and the worshipping of modern technology as well [Bracher, 1976, 546]. In this aspect of the system, mathematicians and scientists found their place; there was hardly room for the reactionary romanticism of ideologues like Bieberbach.

EPILOGUE

During the war Bieberbach appears to have played almost no public role. He was still a member of the leading Nazi group of the Berlin Academy. There he was engaged in the preparation of a mathematical dictionary and in several historical enterprises, while the principal responsibility for the *Jahrbuch* was taken over by Harald Geppert [Grau, 1979]. At the university Bieberbach was still dean of the faculty; Max von Laue, who was certainly one of the most outspoken anti-Nazis among Berlin scientists, reported in an uncensored letter of 1943, "Bieberbach rules the department with commendable objectivity" [Weyl, AS 1b].

After Nazi Germany was defeated Bieberbach was dismissed from his office. After the war he repeated his basic views on styles and types to an American officer who interrogated him, complaining that the Nazis had given too much money for applied mathematics and military research [Grünbaum, AS 1]. He never regained a university position, but kept publishing mathematical articles and books. Some former colleagues and students kept contact and invited him frequently for lectures. Perhaps the first such invitation came from Alexander Ostrowski in Switzerland; he had heard that Bieberbach was earning his living by attending to the central heating of several houses [Ostrowski, AS 1]. On this occasion Bieberbach wrote the apologia (quoted above), which ended with a declaration of regret about "those studies and the connected errors" [Hopf, AS 1].

When I visited Bieberbach shortly before his death, he lived in the house of one of his sons. I met a small, old, and very sick man. His memory was very selective, but he still held that the discrimination between a "German" and an "alien" style had nothing to do with values. Ludwig Bieberbach died on September 1, 1982.

NOTES

[1] The present study is part of a larger research project, funded by the *Stiftung Volkswagenwerk*, on mathematics under National Socialism. I am grateful to Hans Freudenthal, Alexander Ostrowski, Reinhard Siegmund-Schultze, and Renate Tobies who gave helpful comments on a draft of this paper. Special thanks go to Esther Phillips and David Rowe who gave the paper a very careful reading and took pains to eliminate the many linguistic flaws. I am fully responsible for all translations of quotations. The sometimes strange sound of quotations from the thirties is, I hope, not only the result of bad translation, but transmits some of the strangeness the original texts have to present day German readers.

[2] The case of *Deutsche Physik* has, in fact, been the object of careful, if not completely sufficient, historical analysis by A. Beyerchen [1977] and S. Richter

[1980]. The first article on *Deutsche Mathematik* appeared in 1938 and was written by E. J. Gumbel. An explanatory attempt is contained in Booss et al. [1972]. H. Lindner [1980] gives a detailed exposition of Bieberbach's (and Jaensch's) typology. [Quaisser, 1970] was not available to me, but see [Quaisser, 1984]. Besides Beyerchen's book on physics, the most interesting history of a discipline in Nazi Germany is [Geuter, 1984] on psychology.*

[3] Biographical data are not annotated in detail. Aside from obvious sources, like biographical dictionaries, they have been collected from private communications [Bieberbach, AS 1, AS 2a] and from the Bieberbach files in the Berlin Document Center [AS 4] and Humboldt-Universität Berlin [AS 5a].

[4] The book could not be recovered in the archives of Frankfurt University.

[5] This statement of editorial policy is unsigned. However, it appeared after Bieberbach had joined August Gutzmer as editor of the *Jahresbericht*, and a letter from Bieberbach to Klein (March 8, 1921) shows that he was determined to promote a new editorial policy [Klein, AS 1].

[6] A general discussion of the social system of mathematics in relation to National Socialism is given in [Mehrtens, 1985b].

[7] There is no historical account of the events surrounding the Congress. Documents are in the Brouwer archives [AS 1b], the Prandtl papers [AS 1], and in the Hilbert papers [AS 1], where there is a copy of Bieberbach's letter against the Congress as well as the manuscript of Hilbert's opening address to the Congress.

[8] There is no historical account of the "*Annalen*" affair"; the main body of documents is in the Brouwer archives [AS 1c]; Einstein's file on the "Frosch-Mäusekrieg" [AS 1b] contains another collection of the main correspondence concerning the *Annalen*.

[9] "The frog and mouse battle" is the title of a Greek parody of Homer's *Iliad*.

[10] On the *Jahrbuch*, see [Siegmund-Schultze, 1984a]; the reports mentioned are in the archives of the Academy of Sciences of the GDR [*Jahrbuch* AS 1].

[11] Olga Taussky-Todd and Peter Scherk have told the story in the course of informal interviews, October 1984. Both attended the celebration.

[12] On Bieberbach's relation to the *Mathematische Zeitschrift*, I was unable to find any documents. But in 1933 H. Kneser asked Vahlen, who was in the Ministry, whether he should take over the editorship of the *Zeitschrift* which was offered to him by the publishing house Springer. Vahlen answered that he would accept the offer only if the whole board was "*gleichgeschaltet*" (Nazified); the fact that the publishing house was "Jewish" had to be put up with, at least for the moment [Kneser, AS 1a].

[13] The "real" Germans were defined by the Nurnberg laws of 1935. There are very few documents on Bieberbach's relations with the *Compositio*. In July 1934 Gustav Doetsch wrote that he would follow Bieberbach if he resigned from the board, but he was still waiting for the results of a meeting between Bieberbach and Brouwer

*Added in proof: A survey of Bieberbach's mathematical achievements together with a brief biography, a list of publications, and a list of Bieberbach's doctoral students, is given in [Grunsky, 1986].

[Freudenthal, AS 2]. When the first issue of the *Compositio* appeared in 1935, Bieberbach was not named as editor. The three "real" Germans still on the list—Doetsch, Georg Feigl, and Wilhelm Süss—were to leave the board by 1936 [Hopf, AS 1].

[14] The minutes of the Pyrmont assembly were published at considerable length by Bieberbach [DMV, 1934]; further sources on the meeting are the Kneser-Bieberbach correspondence [Kneser AS 1b], the Bohr-Veblen correspondence [Bohr, AS 1], and the Bohr-Courant correspondence [Courant AS 3]. A more detailed description of the Bieberbach–DMV conflict is given in [Mehrtens, 1985a].

[15] The following description of the founding and the fate of *Deutsche Mathematik* is based on a DFG file which is so small that annotation in detail is not necessary [*Deutsche Mathematik*, AS 1]. Further information is taken from the individual issues of the journal (first and last pages).

[16] A more detailed discussion of the typology is given by Lindner [1980]; see also [Quaisser, 1984]. The work and career of Jaensch are discussed by Geuter [1985].

REFERENCES

1. ARCHIVAL SOURCES [AS]

Bieberbach   AS1   Interview with L. Bieberbach by the author, 21. Sept. 1981.

AS2   Papers from the private collection of L. Bieberbach,

a   list of publications

b   typescript "Vom Wissenschaftsideal der Mathematiker, Vortrag gehalten am 15.2.26".

AS3   Collection of reprints and papers of L. Bieberbach, private property, N. Jacob, Erlangen,

a   manuscript, "Vortrag am 3/4 34",

b   correspondence Landau-Bieberbach 1921–25,

c   letter George Polya to Bieberbach, 18. Jan. 1921,

d   letters Wilhelm Blaschke to Bieberbach, 27. Jan. 1921, 6 Mar. 1921.

AS4   File L. Bieberbach, Berlin Document Center.

AS5   Universitätsarchiv der Humboldt-Universität zu Berlin (GDR),

a   Verwaltungsdir., Personalakte, Nr. 220, Ludwig Bieberbach,

b   Philosophische Fakultät, Dekanat, Nr. 1467, Bl. 222, 275–277, Nr. 1468, Bl. 312, Nr. 1469, Bl. 152–3 (statement on Bieberbach), Nr. 1470, Bl. 79.

AS6   Archiv der Akademie der Wissenschaften der DDR, Berlin (GDR), Wahlantrag Bieberbach, AAW II: IIIa, Bd. 23, Bl. 136.

AS7   Archiv des Springer Verlages, Heidelberg, file Bieberbach, B 141, correspondence of Dec. 1925.

Bohr	AS1	Veblen papers, Library of Congress, Washington, D.C., Manuscript Division, Bohr to Veblen, 11. Aug. 1934, 18. Sept. 1934.
Brouwer	AS1	Brouwer Archief, Rijksuniversiteit Utrecht, Mathematical Institute,
	a	correspondence Karl Kerkhof, Brouwer to Kerkhof, 11. May 1925,
	b	file Bologna Congress,
	c	file *Mathematische Annalen*.
Courant	AS1	Archiv des Springer Verlages, Heidelberg, file Courant, C 67
	a	C 67 II, Courant to Springer, 24. Dec. 1925,
	b	C 67 III, Courant to Springer, 1. Jul. 1926.
	AS2	Courant papers, Courant Institute, New York University, file Fraenkel, Courant to Fraenkel, 19. Oct. 1945.
	AS3	Courant papers, private property of Mrs. Courant, New Rochelle, Corresp. H. Bohr 1934–36.
Deutsche Mathematik	AS1	Bundesarchiv Koblenz, R 73, 15 934.
Dingler	AS1	Bundesarchiv Koblenz, NS 12, 806.
	AS2	Hugo Dingler Archiv, Hofbibliothek Aschaffenburg, Faszikel 98, Bieberbach to Dingler, 17. Feb. 1940.
Einstein	AS1	Albert Einstein papers, duplicate archive, Mudd Library, Princeton University,
	a	Box 6, Nr. 6–109
	b	Box 13, Nr. 13–139 to 13–181, Nr. 13–075 (Hilbert to Einstein, 28. Feb. 1925), Nr. 13–156 (note by F. Springer).
Freudenthal	AS1	Interview with Hans Freudenthal by the author, 30. Nov. 1981.
	AS2	Personal files, Hans Freudenthal, Utrecht, copy of letter, Bieberbach to Feigl, 19. Jul. 1934, copy of postcard, Doetsch to Feigl, 17. Jul. 1934.
Grünbaum	AS1	Interview with A. Grünbaum by the author, 14. Oct. 1984.
Hilbert	AS1	Hilbert papers, Staats- und Universitätsbibliothek Göttingen, Handschriftenabteilung, Cod.Ms. Hilbert, 18/1–2.
Hopf	AS1	Heinz Hopf papers, Eidgenössische Technische Hochschule Zürich, Wissenschaftshistorische Sammlungen, Hs 621: 253, Bieberbach to Hopf, 12. Feb. 1949, 335, Brouwer to Hopf, 20. Mar. 1936.
Jahrbuch	AS1	Archiv der Akademie der Wissenschaften der DDR, AAW II: VIIbl, Bd. 7, Heft 1, Nr. 39, 55, 64, 77, 82, 100, 144.
Klein	AS1	Klein papers, Niedersächsische Staats- und Universitätsbibliothek, Göttingen, Handschriftenabteilung, VIII 124 A & B, Bieberbach to Klein 8. Mar. 1921.
Kneser	AS1	Hellmuth Kneser papers, private property Martin Kneser, Göttingen,

	a	Vahlen correspondence, Vahlen to Kneser, 5. Sep. 1933,
	b	Bieberbach correspondence.
von Mises	AS1	Richard von Mises papers, Harvard University, Archives, HUG 4574.5 Box 2, Bieberbach to von Mises, undated (April 1933), HUG 4574.5.2, Box 5, von Mises to his mother, 1. May 1933.
Ostrowski	AS1	Letter by Alexander Ostrowski to the author, 15. Mar. 1985.
Prandtl	AS1	Ludwig Prandtl papers, Max-Planck-Institut für Strömungsforschung, Göttingen, Archiv, file "Prof. Dr. Richard von Mises".
Rado	AS1	Interview with Richard Rado by the author, 21. Mar. 1982.
Veblen	AS1	Letter Veblen to Bieberbach, 19. May 1934, ETH Zürich, Wissenschaftshistorische Sammlungen, Hs 653:1.
Weyl	AS1	Herman Weyl papers, ETH Zürich, Wissenschaftshistorische Sammlungen, Hs.91,
	a	Hs 91:60 Courant to Weyl, 31 May 1934,
	b	Hs 91:658 von Laue to Weyl, 11 Mar. 1943.

2. PUBLICATIONS

Abbreviations

DM	Deutsche Mathematik
MMDDR	Mitteilungen der Mathematischen Gesellschaft der DDR
JbDMV	Jahresbericht der Deutschen Mathematiker-Vereinigung
SbPAW	Sitzungsberichte der Preussischen Akademie der Wissenschaften

Alexandrov, P. S. (ed.), 1971, *Die Hilbertschen Probleme*, Akadem, Verlagsgesellschaft, Leipzig.

Anonymous (signed "P. S."), 1934, "Neue Mathematik—Ein Vortrag von Ludwig Bieberbach," *Deutsche Zukunft* 8, (May 1934) 15.

Beyerchen, A., 1977, *Scientists under Hitler—Politics and the Physics Community in the Third Reich*, Yale Univ. Press, New Haven.

Bieberbach, L., 1910, *Zur Theorie der automorphen Funktionen*, Diss, Göttingen.

_____, 1911/12, "Über die Bewegungsgruppen der Euklidischen Räume," *Math. Ann.* 70, 297–396; 72, 400–12.

_____, 1914, "Über die Grundlagen der modernen Mathematik," *Die Geisteswissenschaften* 1, 896–901.

_____, 1915, *Einführung in die konforme Abbildung*, Göschen, Berlin.

_____, 1916, "Über die Koeffizienten derjenigen Potenzreihen, welche eine schlichte Abbildung des Einheitskreises vermitteln," *SbPAW*, 940–55.

_____, 1917a, *Differentialrechnung*, Teubner, Leipzig.

_____, 1917b, "R. P. Richardson and E. H. Landis, Fundamental Conceptions of Modern Mathematics," (review) *Archiv der Mathematik und Physik* 26, 157–8.

_____, 1918, *Integralrechnung*, Teubner, Leipzig.

_____, 1919, "Zum 70. Geburtstag von Felix Klein," *Frankfurter Zeitung* (23.

238   *Herbert Mehrtens*

Apr. 1919).

_____, 1921a, "Über neuere Lehrbücher der praktischen Analysis," *Zeitschrift für angewandte Mathematik und Mechanik* 1, 61–7.

_____, 1921b, *Lehrbuch der modernen Funktionentheorie I*, Teubner, Leipzig.

_____, 1922a, *Funktionentheorie*, Teubner, Leipzig.

_____, 1922b, "Ziele und Aufgaben des Jahresberichtes der Deutschen Mathematiker-Vereinigung im neuen Jahrgang," *JbDMV* 31, 1–2.

_____, 1922c, "Über Nomographie," *Die Naturwissenschaften* 10, 775–82.

_____, 1922d, "David Hilbert zu sechzigsten Geburtstag," *JbDMV* 31, 3–10.

_____, 1923, *Theorie der Differentialgleichungen*, Springer, Berlin.

_____, 1924a, "Antrittsrede des Hrn. Bieberbach," *SbPAW* XC–XCI.

_____, 1924b, "Über die mathematischen Grundlagen der Nomographie," *VDI Zeitschrift* 68, 495–8.

_____, 1924c, "Ganesh Prasad, Mathematical Research in the Last 20 Years" (review) *Deutsche Literaturzeitung* 45, 725–7.

_____, 1925, "Über die Entwicklung der nichteuklidischen Geometrie im 19. Jahrhundert," *SbPAW* 381–98.

_____, 1933, "Anschauung und Denken in der modernen Mathematik," (abstract) *SbPAW* 643.

_____, 1934a, "Die Kunst des Zitierens - Ein offener Brief an Herrn Harald Bohr in København," *JbDMV* 44, 2.Abt., 1–3.

_____, 1934b, "Persönlichkeitsstruktur und mathematisches Schaffen," *Forschungen und Fortschritte* 10, 235–7.

_____, 1934c, "Persönlichkeitsstruktur und mathematisches Schaffen," *Unterrichtsblätter für Mathematik und Naturwissenschaften* 40, 236–43.

_____, 1934d, "Stilarten mathematischen Schaffens," *SbPAW* 1 351–60.

_____, 1935, "Zweihundertfünfzig Jahre Differentialrechnung," *Zeitschrift für die gesamte Naturwissenschaft* 1, 171–7.

_____, 1936a, (preface, no title) *DM* 1, 3.

_____, 1936b, "Die deutsche Leistung in der Mathematik," *Deutsche Saat in fremder Erde*, K. Römer (ed.), 100–2, Zeitgeschichte Verlag, Berlin.

_____, 1937, "Fragen des mathematischen Universitätsunterrichtes," *DM* 2, 11–6.

_____, 1938a, *Carl Friedrich Gauss—Ein deutsches Gelehrtenleben*, Keil Verlag, Berlin.

_____, 1938b, *Galilei und die Inquisition*, Arbeitsgemeinschaft für Zeitgeschichte, Berlin.

_____, 1940, *Die völkische Verwurzelung der Wissenschaft* (*Typen mathematischen Schaffens*), Sitzungsber. Heidelberger Akad. d. Wiss., Math.-nat. Klasse, Jg. 1940, 5. Abh., Weißsche Universitätsbuchhandlung, Heidelberg.

_____, 1943, "David Hilbert," *Europäischer Wissenschaftsdienst*, 4–5.

Bohr, H., 1934, "'Ny Matematik' i Tyskland," *Berlingske Aften*, (1 May 1934).

Booss, B. et al., 1972, "Gesetzmäßigkeit in der Entwicklung mathematischer Tätigkeit," in *Hegel-Jahrbuch* 1972, W. R. Beyer (ed.), 50–67, A. Hain, Meisenheim.

Boutroux, P., 1927, *Das Wissenschaftsideal der Mathematiker*, Teubner, Leipzig.

Bracher, K. D., 1976, *Die deutsche Diktatur—Entstehung, Struktur, Folgen des Nationalsozialismus*, 5th ed., Kiepenheuer and Witsch, Köln.

Cock, A. G., 1983, "Chauvinism and Internationalism in Science: The International Research Council, 1919–1926," *Notes and Records of the Royal Society* 37, 249–88.

DMV, 1934, "Mitgliederversammlung" (minutes), *JbDMV* 44, 2.Abt., 86–8.

Einstein, A., and M. Born, 1972, *Briefwechsel 1916–1955*, Rowohlt, Reinbek.

Enriques, F., 1927, *Zur Geschichte der Logik*, Teubner, Leipzig.

Forman, P., 1971, "Weimar Culture, Causality, and Quantum Theory, 1918–1927: Adaption by German Physicists and Mathematicians to a Hostile Intellectual Environment," *Historical Studies in the Physical Sciences* 3, 1–115.

———, 1973, "Scientific Internationalism and the Weimar Physicists," *Isis* 64, 151–80.

Fraenkel, A., 1967, *Lebenskreise—Aus den Erinnerungen eines jüdischen Mathematikers*, DVA, Stuttgart.

Geuter, U., 1984, *Die Professionalisierung der deutschen Psychologie im Nationalsozialismus*, Suhrkamp, Frankfurt.

———, 1985, "Nationalsozialistische Ideologie und Psychologie," in *Geschichte der deutschen Psychologie im zwanzigsten Jahrhundert*, M. Ash and U. Geuter (eds.), Westdeutscher Verlag, Opladen.

Grau, C., W. Schlicker, and L. Zeil, 1976, *Die Berliner Akademie der Wissenschaften in der Zeit des Imperialismus, Teil III: Die Jahre der faschistischen Diktatur 1933 bis 1945*, Akademie Verlag, Berlin.

Grunsky, H., 1986, "Ludwig Bieberbach zum Gedächtnis," *JbDMV* 88, 190–205.

Gumbel, E. J., 1938, "Arische Naturwissenschaft?," in *Freie Wissenschaft—Ein Sammelbuch aus der deutschen Emigration*, E. J. Gumbel (ed.), 246–62, Straßburg.

Hardy, G. H., 1934, "The J-type and the S-type among mathematicians," *Nature* 134, 250; repr. *Mathematical Intelligencer* 6 (1984), Nr. 3, 7.

Hasse, H., 1929, "E. Landau, Vorlesungen über Zahlentheorie" (review), *JbDMV* 38, 2. Abt., 52–61.

Heiber, H., 1966, *Walter Frank und sein Reichsinstitut für Deutsche Geschichte*, DVA, Stuttgart.

Jaensch, E. R., 1931, *Über die Grundlagen der menschlichen Erkenntnis*, J. A. Barth, Leipzig.

Jaensch, E. R., and F. Althoff, 1939, *Mathematisches Denken und Seelenform*, J. A. Barth, Leipzig.

Klein, F., 1911, *The Evanston Colloquium Lectures on Mathematics*, Macmillan, New York, 2nd printing.

Kleinert, A., 1979, "Von der Science Allemande zur Deutschen Physik," *Francia —Forschungen zur Westeuropäischen Geschichte* 6, 509–25.

———, 1980, "Lenard, Stark und die Kaiser-Wilhelm-Gesellschaft," *Physikalische Blätter* 36, Nr. 2, 35–43.

———, 1985, "Noch einmal: Sommerfeld und Einstein," *Sudhoffs Archiv*, in print.

Kluge, W., 1983, *Edmund Landau. Sein Werk und sein Einfluß auf die Entwicklung der Mathematik*, Thesis (Staatsexamen), Universität Duisburg.

Lakatos, I., 1968, "Criticism and Methodology of Scientific Research Programmes," *Proceedings Aristotelian Society* 69, 149–86.

Lindner, H., 1980, " 'Deutsche' und 'gegentypische' Mathematik—Zur Begründung einer 'arteigenen Mathematik' im 'Dritten Reich' durch Ludwig Bieberbach," in Mehrtens and Richter, 1980, pp. 88–115.

Manger, E., 1934a, "Felix Klein im Semi-Kürschner!" *JbDMV* 44, 4–11.

_____, 1934b, "Nochmals: Deutsche Mathematik," *Deutsche Zukunft* (13 May 1934) 15.

Mason, T., 1978, *Sozialpolitik im Dritten Reich—Arbeiterklasse und Volksgemeinschaft*, 2nd ed. Westdeutscher Verlag, Opladen.

Mehrtens, H., 1980, "Das 'Dritte Reich' in der Naturwissenschaftsgeschichte: Literaturbericht und Problemskizze," in Mehrtens and Richter, 1980, 15–87.

_____, 1982, "Mathematik als historischer Prozeß: Zum Beispiel die Zeit um 1900," *Beiträge zum Mathematikunterricht* 1982, Schroedel, Hannover, pp. 71–80.

_____, 1984, "Anschauungswelt versus Papierwelt—Zur historischen Interpretation der Grundlagenkrise der Mathematik," in *Ontologie und Wissenschaft*, H. Poser and H. W. Schütt (eds.), Technische Universität, Berlin, 231–276.

_____, 1985a, "Die 'Gleichschaltung' der mathematischen Gesellschaften im nationalsozialistischen Deutschland," *Jahrbuch Überblicke Mathematik*, 83–101.

_____, 1985b, "The Social System of Mathematics and National Socialism, A Survey," *Sociological Inquiry*, in print.

Mehrtens, H., and S. Richter (eds.), 1980, *Naturwissenschaft, Technik und NS-Ideologie*, Suhrkamp, Frankfurt.

Quaisser, E., 1970, *Zum Wirken des Mathematikers Ludwig Bieberbach in der Zeit von 1920 bis 1945*, Thesis, Pädagogische Hochschule "Karl Liebknecht" Potsdam.

_____, 1984, "Zur 'Deutschen Mathematik'," *Wissenschaftliche Zeitschrift der Ernst-Moritz-Arndt-Universität Greifswald, Math.-Nat. Reihe* Jg. 32, Nr. 1–2, 35–9.

Reid, C., 1970, *Hilbert*, Springer, Berlin.

_____, 1976, *Courant in Göttingen and New York—The Story of an Improbable Mathematician*, Springer, New York.

Richter, S., 1980, "Die 'Deutsche Physik'," in Mehrtens and Richter, 1984, pp. 116–141.

Rowe, D., 1984, "Jewish Mathematicians and 'Jewish Mathematics' in the Göttingen Era of Felix Klein," *Isis*, in print.

Schappacher, N., 1984, "Das mathematische Institut 1929–1950," Unpubl. Ms., to appear in *Die Universität Göttingen unter dem Nationalsozialismus*, H. J. Dahms (ed.).

Schröder-Gudehus, B., 1966, *Deutsche Wissenschaft und internationale Zusammenarbeit 1914–1928*, Thèse Université de Genève, Geneva, Imprimerie Dumaret & Golay.

Siegmund-Schultze, R., 1984a, "Das Ende des Jahrbuchs über die Fortschritte der Mathematik und die Brechung des deutschen Referatemonopols," *MMDDR* (1984), Nr. 1, 91–101.

_____, 1984b, "Theodor Vahlen—Zum Schuldanteil eines deutschen Mathematikers am faschistischen Mißbrauch der Wissenschaft," *NTM Schriftenreihe für Geschichte der Naturwissenschaften, Technik und Medizin* 21, Nr.1, 17–32.

Tobies, R., 1981, "Zur Unterstützung mathematischer Forschungen durch die Notgemeinschaft der Deutschen Wissenschaft im Zeitraum der Weimarer Republik," *MMDDR* (1981), Nr.1, 81–99.

Tornier, E., 1936, "Mathematiker oder Jongleur mit Definitionen," *DM* 1, 8–9.

Van der Waerden, B. L., 1930, *Moderne Algebra*, 2 vols., 1930/31, Springer, Berlin.

Weyl, H., 1921, "Über die neue Grundlagenkrise der Mathematik," *Math. Zeitschrift* 10, 39–79.

THE SOLUTION OF PARTIAL DIFFERENTIAL EQUATIONS BY SEPARATION OF VARIABLES: A HISTORICAL SURVEY

Jesper Lützen

INTRODUCTION; FOURIER ON HEAT CONDUCTION

Joseph Fourier was the first to find a complete solution of a partial differential equation using all the ingredients of the method of separation of variables. Therefore I shall present the method by giving an account of the first problem discussed in his "Théorie analytique de la chaleur" [Fourier, 1822, 163–178].[1] It deals with the stationary temperature distribution in a semi-infinite rectangular plate (Figure 1).

Fourier had discovered that the temperature $u(x, y)$ must satisfy Laplace's equation,

$$\frac{\partial^2 u}{\partial x^2} + \frac{\partial^2 u}{\partial y^2} = 0. \tag{1}$$

---

[1]Fourier had presented this investigation to the Académie des Sciences in his memoire of 1807 [Grattan-Guinness, 1972 pp. 131-172].

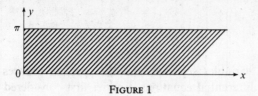

FIGURE 1

He assumed that the temperature at the horizontal infinite boundaries are maintained at zero degrees, so that

$$u(x,0) = u(x,\pi) = 0, \tag{2}$$

and for physical reasons he also argued that

$$u(x, y) \to 0 \quad \text{as } x \to \infty. \tag{3}$$

Finally, he assumed that a source of heat maintained the left boundary at a fixed temperature $u_0(y)$, that is,

$$u(0, y) = u_0(y) \quad \text{for } 0 < y < \pi. \tag{4}$$

He first looked for solutions to (1) that can be written as products of a function of x and a function of y:

$$u(x, y) = F(x)f(y). \tag{5}$$

In this case equation (1) is reduced to $F''(x)f(y) + F(x)f''(y) = 0$ or, equivalently,

$$\frac{F''(x)}{F(x)} = -\frac{f''(y)}{f(y)}. \tag{6}$$

Fourier argued that since the left-hand side of (6) is only a function of x and the right-hand side only a function of y, they can be equal only if each is equal to a constant, say m^2, so that

$$F''(x) = m^2 F(x) \tag{7}$$

and

$$f''(y) = -m^2 f(y). \tag{8}$$

Thus the variables separate, and the problem reduces to one in ordinary differential equations. Fourier first considered (8), whose general solution is

$$f(y) = A \sin my + B \cos my. \tag{9}$$

According to (2) f must also satisfy the boundary conditions

$$f(0) = f(\pi) = 0, \tag{10}$$

which implies that m is an integer, and that

$$f(y) = A \sin my, \qquad m \in \mathbf{Z}. \tag{11}$$

In modern terminology we would say that the eigenvalue problem, (8) and (10), has eigenvalues m in $\mathbf{Z}$ and that the eigenfunctions are given by (11). The corresponding solutions to (7) are $e^{\pm mx}$. Taking (3) and the symmetry of $\sin my$ into account, Fourier found the following particular solutions to (1)–(3):

$$u(x, y) = A e^{-mx} \sin my, \quad m = 0, 1, 2, \ldots. \tag{12}$$

Because of the linearity of the equations (1)–(3), the infinite linear combination

$$u(x, y) = \sum_{m=1}^{\infty} A_m e^{-mx} \sin my \tag{13}$$

is also a solution.[2]

---

[2] In fact, in this example Fourier assumed that the boundaries of the plate were at $y = -\pi/2$ and $y = \pi/2$, and he assumed $u_0(y) = 1$ from the start, and therefore only considered the eigenfunctions $\cos mx$ where m is uneven.

Fourier was then faced with the problem of determining coefficients A_m such that (4) is fulfilled, that is

$$u_0(y) = \sum_{m=1}^{\infty} A_m \sin my, \quad \text{for } 0 < y < \pi. \tag{14}$$

At first he used a lengthy argument to find the A_m when $u_0(y) = 1$. However, later in the book he simply multiplied equation (14) by $\sin ny$ and integrated both sides from 0 to π:

$$\int_0^\pi u_0(y) \sin ny \, dy = \int_0^\pi \sum_{m=1}^{\infty} A_m \sin my \sin ny \, dy. \tag{15}$$

Interchanging the order of integration and summation, Fourier noted that

$$\int_0^\pi \sin ny \sin my = \begin{cases} 0 & \text{for } m \neq n \\ \dfrac{\pi}{2} & \text{for } m = n, \end{cases} \tag{16}$$

from which the formulas for the so-called Fourier coefficients follows:

$$A_m = \frac{2}{\pi} \int_0^\pi u_0(y) \sin my \, dy. \tag{17}$$

The formulas for Fourier coefficients for the cosine series, and for the full Fourier series on the interval $[-\pi, \pi]$ were found in a similar manner.[3]

Fourier was in fact only able to show that if a function $u_0(y)$ can be expressed as a sine series (14) (and if one can interchange integration and summation), then the coefficients are determined by (17). Fourier also tried, with several examples and insufficient arguments, to convince his readers that if u_0 is an arbitrary

---

[3] Of course the series (13) must be convergent and it must be permissible to differentiate $u(x, y)$ by differentiating the series term by term. Fourier discussed convergence later, but he never realized the last problem.

function and if A_n is determined by (17), then the Fourier series (14) does indeed represent the function $u_0(y)$. However, another seven years were to pass before Dirichlet correctly proved that under suitable conditions the Fourier series converges to the expanded function and that therefore the solution (13), found by separation of variables, is indeed the general solution.

The procedure employed by Fourier is an extremely powerful method for solving partial differential equations. It consists of a series of important components which include

1. the search for product solutions (5) that separate the variables;

2. the solution of the eigenvalue problem (8) and (10); (It is particularly important to note that there is only a discrete spectrum of eigenvalues, all of which are real.)

3. the idea that the special solutions (12) can be superimposed;

4. a method whose generality is secured by the fact that sufficiently general functions can be expanded in Fourier series; an important step toward establishing this fact is

5. the determination of the formulas (17) for the Fourier coefficients. These, in turn are consequences of

6. the orthogonality relations.

The development and extensions of these ideas are subjects of this paper.

The emergence of mathematical ideas usually defies the logical principles that characterize the ideas themselves, and the method described above is no exception. For example, it is striking that many of the central ideas in the method were discovered before any partial differential equation had been formulated; how that happened in the period 1710–50, will be described in the first section. The second section is devoted to some important applications of the separation of variables in mathematical physics during the century following the discovery of the method. They led to many *special* eigenvalue problems. Finally, separation of variables also raised problems concerning the spectral theory for the *general* second-order self-adjoint linear differential equations. In the last section we discuss the pioneering work on this subject carried out by Sturm and Liouville.

I. THE EMERGENCE OF THE FUNDAMENTAL IDEAS

The central ideas listed above emerged in the framework of the vibratory motion of continuous media with its close link to the theory of music. Eigenfunctions have their origins in the simple modes of a vibrating string which were investigated in the 17th century by John Wallis [1677] and Francis Robartes [1692] who described the connection between the number of nodes and the overtones. Early in the 18th century Joseph Sauveur and Bernard le Bovier Fontenelle explained that, by assuming that a string can vibrate with many simple modes at the same time, overtones can be emitted simultaneously with the fundamental tone. This is the musico-physical origin of the principle of superposition. However, it took half a century before these ideas were incorporated into the mathematico-mechanical theory of vibratory motion.

Thus in the first mathematical treatment of the vibrating string, published by Brook Taylor [1713 and 1715], the higher modes were completely overlooked. Taylor began by proving that the accelerating force of a point on the string is proportional to the curvature of the string at that point. If a horizontal axis (x) is determined by the fixed points A and B of the string, and if the displacements are called y (see Figure 2) the modern reader would immediately observe that the accelerating force is proportional to $\partial^2 y / \partial t^2$ and the curvature is proportional to $\partial^2 y / \partial x^2$, so that we have the wave equation

$$\frac{\partial^2 y}{\partial t^2} = c^2 \frac{\partial^2 y}{\partial x^2}. \tag{18}$$

Taylor, however, made neither of these observations. Instead he limited his investigations to vibrations for which all the points on the string cross the x-axis simultaneously and concluded, incor-

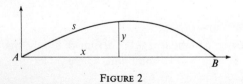

FIGURE 2

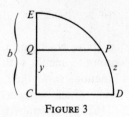

FIGURE 3

rectly, that the curvature at a point is proportional to the distance y of that point from the axis. He then used the expression $\ddot{x}/\dot{s}\dot{y}$ for the curvature (the dots mean differentiation with respect to the arclength S) and found the differential equation

$$y = a^2\ddot{x}/\dot{s}\dot{y}. \tag{19}$$

Taylor integrated (19) once to find, in the limit where the magnitude of the vibrations becomes small, that

$$\dot{x} = \frac{a\dot{y}}{\sqrt{b^2 - y^2}}, \tag{20}$$

where b is the maximum value of y. In order to solve equation (20) he drew a circle quadrant ED (see Figure 3) with radius b; at the distance $y = CQ$ from the radius CD he drew QP and defined z to be equal to the arc DP. It was well known that

$$\dot{z} = \frac{b\dot{y}}{\sqrt{b^2 - y^2}}, \tag{21}$$

so that from (20) $\dot{z} = b\dot{x}/a$ follows or, neglecting a constant of integration, $z = bx/a$. In contemporary notation,

$$y = b \sin\frac{z}{b} = b \sin\frac{x}{a}. \tag{22}$$

Without further comment Taylor assumed finally that the string obtains its maximum at its midpoint, where he let z be the quarter arc of the circle ED. Thus if l denotes the length of the string $z = b(\pi/2) = (b/a)(l/2)$ from which $a = l/\pi$ follows, so that in modern notation we have

$$y = b \sin(x\pi/l). \tag{23}$$

Taylor argued that if the curve is initially of this form it will so remain, and since the acceleration is proportional to the distance from the axis each point will oscillate like a pendulum. From the value $a = l/\pi$ he was finally able to determine the length of the pendulum with the same period. In this way he derived the law for the frequency of the vibrations experimentally derived almost a century earlier by Marin Mersenne.

There are two remarkable missed opportunities in Taylor's paper. First, by assuming the curvature to be proportional to the displacement, he missed deriving the wave equation. That is, he separated the variables and obtained a differential equation (19) showing how y depends on x and, implicitly, another for the variation with respect to t. The latter, however, he immediately related to the equation of the pendulum. This unconscious physical separation of variables is in fact equivalent to assuming $y(x, t) = f(x)F(t)$. Secondly, Taylor missed the higher modes, which is astonishing for a person with his great interest in music. In the construction corresponding to (22) he had all the simple modes, but by taking $z = ED$ at the midpoint of the string, he excluded the higher modes. His neglect may be due to a combination of two reasons. First, he aimed at a proof of Mersenne's law that deals only with the fundamental mode. Second, as pointed out by Cannon and Dostrovsky [1981], the extension of the sine function to more than half a period is less apparent when it is defined geometrically (as in Figure 3) than in the modern definition. In fact, higher modes first appeared in a mathematical treatment when they were described by Daniel Bernoulli [1733, 1734] in analytic rather than geometrical terms.

In his papers on the vibrations of a hanging chain, D. Bernoulli approached the problem with a method borrowed from his father, Johann, who had applied it to clarify Taylor's analysis [J. Bernoulli, 1727]. The idea is to investigate a weightless string loaded with finitely many equidistant point-masses, whose number tend to infinity while their total mass is kept constant. D. Bernoulli first considered two point masses, one situated below the other at a distance equal to the distance from the top mass to the point of suspension. He proved that this system has two simple modes whose periods he determined. Similarly, if the weightless string is loaded with three point masses, there will be three simple modes, and so on.

D. Bernoulli derived these results by finding the accelerating force on each point mass as a function of its position, equating it, as his father and Taylor had done, to a constant times its displacement in order "to obtain isochronism." In the continuous limit, where there are infinitely many infinitely small point masses, the expression for the accelerating force becomes

$$\frac{dy}{dx} + x\frac{d^2y}{dx^2},$$ (24)

where y is the displacement and x the distance from the bottom of the chain. Thus the equation for the simple modes is

$$\frac{dy}{dx} + x\frac{d^2y}{dx^2} + \frac{y}{\nu} = 0.$$ (25)

Bernoulli expressed the solution of the equation as a power series which we recognize to be equivalent to

$$y = J_0\left(2\sqrt{x/\nu}\right),$$ (26)

where J_0 is the zeroth Bessel function. When the chain is fixed at its top ν must be a root of the equation

$$J_0\left(2\sqrt{l/\nu}\right) = 0,$$ (27)

where l is the length of the string. He remarked, without proof, that there are infinitely many such (eigen)values ν and that to each of these corresponds a simple mode (26) whose period is proportional to $\nu^{-1/2}$.

Bernoulli did not realize that by discovering the higher modes he had made an advance over his predecessors. Indeed, in a scholium he remarked,

> Moreover, in a chain that can be taken as though infinitely long the [shape of the] highest arc does not differ from the shape of a tense musical string.... Nor is it difficult from this theory to deduce a theory of musical strings which clearly agrees with those that Taylor and my father gave. [D. Bernoulli, 1733, 119][4]

In a later paper D. Bernoulli [1740] described in words, without mathematical formalism, the higher modes of a musical string and their relations to overtones; two years later in the course of investigating yet another vibrating system, namely the rod, he stated the principle of superposition of simple modes.

In 1734 D. Bernoulli had shown that the displacement y of the simple modes of an elastic rod is described by the differential equation

$$\frac{d^4 y}{dx^4} = ky \tag{28}$$

(cf. [Truesdell, 1960]). At first neither he nor his friend Leonhard Euler could solve (28) except by using series, but in 1739 Euler found the exponential solutions to the equation and, more generally, to any linear differential equation with constant coefficients. Euler remarked on this solution that, due to the linearity of the equation, a linear combination of solutions is again a solution and that the general solution can be obtained in this way. In the published paper Euler [1743] showed also how to treat multiple roots.

With these solutions at hand Bernoulli [1742] described the sound emitted from the rod, remarking about the simple modes,

> I have said that often in this experiment both sounds exist together and are perceived, nor is it any wonder since neither oscillation helps or hinders the other.

---

[4] Translation [Cannon and Dostrovsky, 1981, 163].

> Indeed when the band is curved by reason of one oscillation, it may always be considered as straight in respect to another oscillation, since the oscillations are virtually infinitely small. [D. Bernoulli, 1742, Section 8][5]

This formulation of the principle of superposition does not differ from the one given by Fontenelle, but with Bernoulli it acquired a permanent place in mechanics. However, the mathematical meaning of the principle remained obscure since, although each of the simple modes is a solution to an ordinary differential equation (28) for a specific value of k, a linear combination of simple modes does not satisfy the equation for any value of k. Thus there was no connection between Euler's theorem on superposition of solutions to linear differential equations and Bernoulli's superposition principle. The link was not established until the discovery by Jean le Rond d'Alembert of the partial differential equation of motion in the *Traité de Dynamique* [1743].

Having recalled D. Bernoulli's expression (24) for the accelerating force of a point on the hanging chain, d'Alembert observed that if the vibrations of the curve are not simple "it will change in its equation from one instant to the next, and the general value of an ordinate y can be expressed only by a function of $[x]$... and of the time t. In general, then, let $y = \varphi(t, [x])$," [d'Alembert, 1743, 116–117].[6]

Instead of equating the accelerating force to a multiple of the displacement, he equated it to $\partial^2 y / \partial t^2$, thus obtaining the partial differential equation

$$\frac{\partial^2 y}{\partial t^2} = \frac{\partial y}{\partial x} + x \frac{\partial^2 y}{\partial x^2}. \tag{29}$$

This is such an obvious step for us, that one may wonder why it took so long before it was taken. Lack of understanding of the mechanical principles may have been a reason; in particular it was not clear at that time that what we call Newton's second law was the basic equation of mechanics. Moreover, there were mathemati-

---

[5] Translation [Truesdell, 1969, 197].
[6] Translation [Truesdell, 1960, 192].

cal obstacles to the understanding of problems with more than one independent variable, and partial derivatives had only gradually emerged during the preceding years in connection with problems of families of curves [Engelsman, 1984]. In fact, (29) was the first proper partial differential equation in the history of mathematics, and at first it remained unnoticed, probably because d'Alembert could do nothing with it. But three years later it created much excitement when d'Alembert [1746] derived the wave equation

$$\frac{\partial^2 y}{\partial t^2} = \frac{\partial^2 y}{\partial x^2} \qquad (30)$$

for the vibrating string and found its solutions,

$$y(x, t) = \psi(t + x) + \varphi(t - x), \qquad (31)$$

using the techniques developed by Euler and others to treat partial derivatives. Here ψ and φ are arbitrary functions. D'Alembert further observed that if $y(x, 0) = 0$ and if the string has an initial velocity which will give it the shape of a sine curve, then

$$y(x, t) = -\frac{1}{2} A \big(\cos n(t + x) - \cos n(t - x) \big) \qquad (32)$$

$$= A \sin nx \sin nt.$$

Here for the first time is the expression for the displacement of a simple mode as a function of space and time. D'Alembert remarked that this function can be expressed as a product $y(x, t) = F(x) f(t)$. Conversely he argued that if a solution to the wave equation is of this form, it must be one of the fundamental modes. He first gave a rather long winded "mechanical" proof, but in a subsequent treatment [d'Alembert, 1750] he remarked that if the equation

$$\psi(t + x) + \varphi(t - x) = F(x) f(t) \qquad (33)$$

is differentiated twice with respect to t and twice with respect to x, one obtains

$$\frac{1}{f(t)} \frac{d^2 f(t)}{dt^2} = \frac{1}{F(x)} \frac{d^2 F(x)}{dx^2}. \qquad (34)$$

"These quantities," he remarked, "must not only be equal but even identical, that is to say they must be equal to the same quantity independently of any equation between t and $[x]$." Therefore $d^2f(t) = m^2dt^2f(t)$ and $d^2F(x) = m^2dx^2F(x)$, from which he deduced the result. This is the first application of the method of separating variables. However, it is interesting to note that d'Alembert did not apply it directly to the partial differential equation (30), but made a detour around the functional equation (33).

D'Alembert did not discuss superposition of simple modes, but soon thereafter Euler did so in his reaction to d'Alembert's paper [Euler, 1748]. The main objective of Euler's paper was to generalize d'Alembert's solution. D'Alembert had required that ψ and φ in (31) be analytic expressions, while Euler, in this and later papers, tried to show that these functions may correspond to arbitrary hand-drawn curves. This gave rise to a long controversy that sheds light on the function concept of that time (cf. [Ravetz, 1961], [Youschkevich, 1976], and [Lützen, 1983]). However, for the present discussion it is of more interest that at the end of his paper Euler gave a "special example" of a motion of the string for which the curve was given by an analytical expression,

$$y = \sum A_n \sin(n\pi x/l)\cos(n\pi t/l), \qquad (35)$$

where l is the length of the string. This is a superposition of simple modes. Thus Euler had shown that the earlier musico-mechanical principle of superposition could be traced back to a mathematical theorem about linear combinations of solutions to linear partial differential equations.

By 1750 all but one of the major constituents of the method of separation of variables had been discovered; it remained unclear how general solutions were obtained in this way. The question was first raised by D. Bernoulli [1753] who, having seen d'Alembert's and Euler's entirely new approach, tried to clarify the ideas on the vibrating string that he himself had hinted at in earlier papers on the hanging chain and the vibrating rod. As in his earlier papers D. Bernoulli treated the x and t variations separately, so that applying the principle of superposition to the simple modes, he arrived only at (35) for $t = 0$:

$$y = \sum_{n=1}^{\infty} A_n \sin(n\pi x/l).$$ (36)

So one sees this infinity of curves found without calculations, and our equation is the same as M. Euler's. It is true that M. Euler does not consider this infinitely infinite multitude to be general, and that in §30 he mentions it only as a special case; but on this point I have not yet sufficiently clarified things in my own mind. If there are other curves, I do not understand in which sense one can allow them.   [D. Bernoulli, 1753, p. 157][7]

The question was later put more clearly by Euler: can any initial curve $f(x)$ be developed in a trigonometric series (36)? Bernoulli convinced himself and tried to convince others that this could indeed be done by determining, in a convenient manner, the infinitely many coefficients A_n. However, he displayed no means of finding the coefficients. Euler and d'Alembert, on the other hand, believed that arbitrary functions could not be represented by trigonometric series. For example, Euler [1753, §9] observed that an expression of the form (36) is odd and periodic with period $2l$ and concluded: "but no algebraic curve can have these properties, so they must all be excluded from the equation."

Here Euler made one of two mistakes due to his algebraic concept of function. Either he did not understand that it makes no sense to talk about the periodicity and parity of an initial curve $f(x)$ defined only in the interval $[0, l]$, or he thought that the initial curve cannot be of the form (36) if its analytic continuation is not globally of this form. The latter argument overlooks the fact that the trigonometric series might represent the function only in the relevant interval $(0, l)$. Although D. Bernoulli later became aware of this fact, most mathematicians were initially convinced by Euler's arguments, so that the method of separation of variables was believed to yield only special solutions.

Generally this belief was held until Fourier's time, despite the prior discovery of the formulas for the Fourier coefficients in

---

[7]My translation.

various astronomical connections. Jean Baptiste Clairaut had found such formulas [Clairaut, 1757], and later they were derived by Euler [1777] using the orthogonality relations just as Fourier [1822] did. To be sure the formulas only show how to determine the coefficients if the function is known to have a trigonometric expansion. Whether a function did indeed have a trigonometric expansion remained a matter of belief, and even Fourier was able to give only inadequate arguments. The question was settled by Peter Gustav Lejeune Dirichlet in a famous paper [Dirichlet, 1829] in which he proved that the Fourier series of an arbitrary, piecewise continuous, piecewise monotonic function converges to the function in the interval $(0, 2\pi)$ (except at points of discontinuity).

Thus it was established that in many special cases the method of separation of variables led to the general solution.

II. SPECIAL EQUATIONS, SPECIAL FUNCTIONS

The partial differential equations of motion of mathematical physics are most often of the second order, and when their coefficients are constant, the eigenfunctions of the separated equations are typically trigonometric functions; the problems of the vibrating homogeneous string and heat conduction in a homogeneous plate (discussed above) are early examples. However, even in the eighteenth century mathematicians had encountered equations with variable coefficients and were thus led to more complicated eigenvalue problems. These equations, which originated in two ways, are illustrated by two examples taken from Euler.

Although Euler doubted the generality of the method of separation of variables, he often used it when he could not find the solutions in closed form. For example, he fell back on this method when, for the first time [Euler, 1759], he studied the motion of a two-dimensional body, namely a membrane. Thus having set up the wave equation

$$\frac{1}{c^2} \frac{\partial^2 z}{\partial t^2} = \frac{\partial^2 z}{\partial x^2} + \frac{\partial^2 z}{\partial y^2}, \tag{37}$$

he separated the variables and obtained the simple modes as a product of three sine functions. Superposition of these solutions was well adapted to the rectangular drum, with boundary conditions on lines parallel to the x- and y-axes, but in the case of a circular drum this method does not work. Here Euler used polar coordinates, r and φ, to transform (37) into an equation with variable coefficients

$$\frac{1}{c^2} \frac{\partial^2 z}{\partial t^2} = \frac{1}{r} \frac{\partial z}{\partial r} + \frac{\partial^2 z}{\partial r^2} + \frac{1}{r^2} \frac{\partial^2 z}{\partial \varphi^2}. \tag{38}$$

Since the coefficients depend only on r, Euler knew how a product solution $z(t, r, \varphi) = u(t)v(r)w(\varphi)$ must depend on t and φ; so he immediately set out to search for solutions of the form $z(t, r, \varphi) = u(r)\sin(\alpha t + a)\sin(\beta \varphi + b)$. He found that u must satisfy the equation

$$u'' + \frac{1}{r} u' + \left(\frac{\alpha^2}{c^2} - \frac{\beta^2}{r^2} \right) u = 0 \tag{39}$$

(later called Bessel's equation) and determined the power series expansion of its solution $u(r) = J_\beta(\alpha r/c)$. Since z must be periodic with period 2π in φ, it follows that β must be an integer, and if the boundary $r = r_0$ is kept fixed α must satisfy the boundary condition,

$$J_\beta(\alpha r_0/c) = 0. \tag{40}$$

Euler claimed correctly, but without proof, that this equation has an infinity of solutions, so that there is a double infinity of simple modes. He knew that the second-order equation (39) ought to have two independent solutions, but his attempt to find the second solution was wrong. For $\beta = 0$, however, he later gave the solution Y_0 in a paper dealing with the vibrations of a hanging cord. This takes us to the second example of a problem leading to a differential equation with variable coefficients.

D'Alembert had already found the equation describing the motions of a freely hanging chain (29) [d'Alembert, 1743].

Euler [1774a] applied the separation of variables to this equation and found the solution as a superposition of the simple modes:

$$J_0(2\omega\sqrt{x})\sin(\omega t + \zeta), \qquad (41)$$

where ω must satisfy the boundary condition

$$J_0(2\omega\sqrt{l}) = 0. \qquad (42)$$

The dependence on x agreed with Daniel Bernoulli's earlier result (26).

In another paper of the same year [1774b], Euler similarly investigated a vertically suspended heavy taut string, and it was in this connection he found Y_0.[8]

These examples illustrate the two different ways in which variable coefficients were introduced. In the latter type the physical quantities describing the problem vary. In this example it was the tension of the string that varies, while in many other examples quantities describing the material may vary. For example, Euler, D. Bernoulli, and d'Alembert discussed the vibration of cords of variable thickness. The general problems raised in this connection led to the Sturm-Liouville theory, to be discussed in the next section. The example of the vibrating drum shows that even when the physical quantities are constant, variable coefficients may arise when there is a change of variables made to suit the boundary conditions. In the remainder of this section I shall cite some of the most important eigenvalue problems with variable coefficients which arose in this way.

One such class of problems was encountered in celestial mechanics, more specifically in the potential theory for celestial bodies. These problems were brought into the realm of partial differential equations when Pierre Simon de Laplace [1782] formulated Laplace's equation to be satisfied by the potential V. The spheroidal

---

[8] The other Bessel functions of the second kind were found a century later by C. G. Neumann and Hankel.

shape of the planets suggested the use of spherical coordinates r, θ and φ ($\mu = \cos \theta$), so he immediately wrote down his equation in these coordinates:

$$\frac{\partial}{\partial \mu}\left((1 - \mu^2)\frac{\partial V}{\partial \mu}\right) + \frac{1}{1 - \mu^2}\frac{\partial^2 V}{\partial \phi^2} + r\frac{\partial^2(rV)}{\partial r^2} = 0. \quad (43)$$

Laplace sought solutions of the form

$$V = \sum_{n=0}^{\infty} \frac{Y_n(\theta, \phi)}{r^n} \quad (44)$$

and found by substitution into (43) that each Y_n must satisfy the equation

$$\frac{\partial}{\partial \mu}\left[(1 - \mu^2)\frac{\partial Y_n}{\partial \mu}\right] + \frac{1}{1 - \mu^2}\frac{\partial^2 Y_n}{\partial \phi^2} + n(n+1)Y_n = 0. \quad (45)$$

The solutions Y_n to this equation are now called the spherical harmonics. Laplace studied them in detail and proved an orthogonality relation for them:

$$\int_{-1}^{1}\int_{0}^{2\pi} Y_n(\mu, \phi)Y_m(\mu, \phi)\,d\mu\,d\phi = 0 \quad \text{for } m \neq n. \quad (46)$$

He also gave an (insufficient) proof, that arbitrary functions on the unit ball can be expanded in a series of spherical harmonics.

Having seen Laplace's paper, Adrien Marie Legendre [1784] proved similar theorems for the Legendre polynomials, which he had introduced in his earlier investigations of the same problem. These polynomials are spherical harmonics that are independent of ϕ.

Polar and spherical coordinates are special cases of curvilinear coordinates. The use of more general systems of curvilinear coordinates in the theory of partial differential equations was studied in depth by Gabriel Lamé in the 1830s. He gave an explicit formulation of an idea, which had guided Euler, Laplace, and his other predecessors, namely: when a boundary value problem is solved by separation of variables, it is desirable that one of the coordinates be constant on the boundary. Thus when a boundary problem is given, one must look for such a system of coordinates that in addition permits a convenient transformation of the equation. In the case of Laplace's equation, Lamé saw that the coordinate system must be orthogonal; that is, if the coordinates are called ρ, μ and ν, any three surfaces $\rho = \rho_0$, $\mu = \mu_0$ and $\nu = \nu_0$ intersect each other orthogonally.

Lamé employed this general method to solve the problem of a stationary temperature distribution in an ellipsoid [Lamé, 1833, 1837, and 1839]. The general procedure is illustrated by sketching Lamé's approach to this natural continuation of Fourier's investigations of heat conduction in a rectangular plate, a cylinder and a sphere. Assuming that the ellipsoid has the equation

$$\frac{x^2}{\rho_0^2} + \frac{y^2}{\rho_0^2 - b^2} + \frac{z^2}{\rho_0^2 - c^2} = 1, \tag{47}$$

where $b^2 < c^2 < \rho_0^2$ Lamé introduced the ellipsoidal coordinates (ρ, μ, ν) of a point having rectangular coordinates (x, y, z) as the solution to the equations,

$$\frac{x^2}{\rho^2} + \frac{y^2}{\rho^2 - b^2} + \frac{z^2}{\rho^2 - c^2} = 1$$

$$\frac{x^2}{\mu^2} + \frac{y^2}{\mu^2 - b^2} - \frac{z^2}{c^2 - \mu^2} = 1 \tag{48}$$

$$\frac{x^2}{\nu^2} - \frac{y^2}{b^2 - \nu^2} - \frac{z^2}{c^2 - \nu^2} = 1,$$

where $0 < \nu^2 < b^2 < \mu^2 < c^2 < \rho^2$. He showed that this is an orthogonal coordinate system, for which it is readily seen that the equation of the boundary ellipsoid is $\rho = \rho_0$. The steady state temperature V must satisfy the potential equation, which in the new coordinates has the form

$$\left(\mu^2 - \nu^2\right)\frac{\partial^2 V}{\partial \epsilon_\rho^2} + \left(\rho^2 - \nu^2\right)\frac{\partial^2 V}{\partial \epsilon_\mu^2} + \left(\rho^2 - \mu^2\right)\frac{\partial^2 V}{\partial \epsilon_\nu^2} = 0, \tag{49}$$

where ϵ_ρ is determined by the elliptic integral,

$$\epsilon_\rho = \int_0^\rho \frac{d\rho}{\sqrt{\rho^2 - b^2}\,\sqrt{\rho^2 - c^2}}, \tag{50}$$

and ϵ_μ, ϵ_ν are defined in a similar way. Lamé sought solutions to (49) of the form

$$V(\rho, \mu, \nu) = R(\rho)M(\mu)N(\nu) \tag{51}$$

and found that R must be a solution of the differential equation,

$$(\rho^2 - b^2)(\rho^2 - c^2)\frac{d^2R}{d\rho^2} + (2\rho^3 - (b^2 + c^2)\rho)\frac{dR}{d\rho}$$

$$= (n(n+1)\rho^2 - B)R, \tag{52}$$

where $n = 0, 1, 2 \ldots$. Similar equations must be satisfied by M and N. He observed that for each value of n, there is a finite number of values of B for which there is a solution R that is either a polynomial in ρ^2, or such a polynomial multiplied by one, two, or three of the factors ρ, $\sqrt{\rho^2 - b^2}$, $\sqrt{\rho^2 - c^2}$. These polynomials are called Lamé functions of the first kind. Other solutions, called Lamé functions of the second kind, were introduced in 1842 by Joseph Liouville (published in [1845]) in connection with his researches on the shape of rotating fluid ellipsoids, and independently by Heinrich Eduard Heine (cf. [Lützen, 1984b, 136 and 148]).

Lamé proved the orthogonality relations

$$\int\int hM_iN_i\,M_jN_j\,d\bar\sigma = 0, \tag{53}$$

where the integral ranges over the surface $\rho = \rho_0$; M_iN_i and M_jN_j correspond to different pairs of (n, B) in (40), and

$$h = h(\rho, \mu, \nu) = \frac{\sqrt{\rho^2 - b^2}\,\sqrt{\rho^2 - c^2}}{\sqrt{\rho^2 - \mu^2}\,\sqrt{\rho^2 - \nu^2}}. \tag{54}$$

This helped Lamé to determine the constants A_{nB} in the general solution, $V = \sum_{n,B} A_{nB} R_{nB} M_{nB} N_{nB}$, when a value on the surface $\rho = \rho_0$ was prescribed. The completeness of the orthogonal system $M_{nB} N_{nB}$ was established by Liouville [1846] using Dirichlet's corresponding result for the spherical harmonics [Dirichlet, 1837].

Lamé introduced several other curvilinear coordinate systems, most of them degenerate cases of the ellipsoidal coordinates. He conducted a thorough study of their geometric properties, and wrote a book on the subject [Lamé, 1859].

After Lamé, the separation of variables in curvilinear coordinates became a standard procedure for the solution of partial differential equations. Many eigenvalue problems for ordinary differential equations originated in the procedure: solutions of these equations, often called special functions, were studied in great detail, in particular with respect to orthogonality relations and completeness. For example, Emile Léonard Mathieu [1868] studied the wave equation in elliptic cylindrical coordinates and was led to what we now call Mathieu functions. The following year Heinrich Weber used parabolic coordinates with a similar purpose in mind and discovered the functions named after him.

Thus the method of separation of variables, when applied to the partial differential equations of physically interesting phenomena and expressed in various curvilinear coordinates, gave rise to a host of ordinary differential equations. The study of their "special" solutions formed the subject of a considerable number of mathematical investigations throughout the 19th century. Some of this work was theoretically interesting, and most of it was useful for applications.

III. STURM-LIOUVILLE THEORY

As already indicated, questions of vibrations and heat conduction in heterogeneous media gave rise to the Sturm-Liouville theory. Charles Francois Sturm [1836b] described as his motivation heat conduction in an inhomogeneous bar with an unequally polished surface. Poisson had shown that the temperature of such a bar is governed by the equation

$$g\frac{\partial u}{\partial t} = \frac{\partial(k\partial u/\partial x)}{\partial x} - lu, \tag{56}$$

where $u(x, t)$ denotes the temperature at time t at the point x, and g, k, and l are positive functions of x. If the temperature of the surrounding medium is maintained at zero degrees, then u must satisfy the following conditions at the endpoints of the bar:

$$k\frac{\partial u}{\partial x} - hu = 0 \qquad \text{for } x = \alpha \tag{57}$$

$$k\frac{\partial u}{\partial x} + Hu = 0 \qquad \text{for } x = \beta; \tag{58}$$

h and H are positive constants. Often the initial temperature is known,

$$u(x,0) = f(x). \tag{59}$$

Sturm separated the variables by setting $u(x, t) = V(x)e^{-rt}$. From (56)–(58) he concluded that V must be a solution to the boundary value problem

$$(k(x)V'(x))' + (g(x)r - l(x))V(x) = 0 \tag{60}$$
$$\text{for } x \in (\alpha, \beta)$$

and

$$k(x)V'(x) - hV(x) = 0 \qquad \text{for } x = \alpha, \tag{61}$$

$$k(x)V'(x) + HV(x) = 0 \qquad \text{for } x = \beta. \tag{62}$$

The boundary value problem (60)–(62) was the one studied by

Sturm and Liouville. More particularly, their investigations concerned

1. properties of the eigenvalues, for example that there are infinitely many $r_1, r_2, r_3, \ldots$;

2. qualitative behavior of the corresponding eigenfunctions $V_1, V_2, V_3, \ldots$;

3. expansion of arbitrary functions in an infinite series of eigenfunctions.

The expansion problem arose in attempting to fit the solution $\sum A_n V_n(x) e^{-r_n t}$, found by superposing simple solutions, to the initial condition (59), that is by setting

$$\sum A_n V_n(x) = f(x). \tag{63}$$

Sturm investigated 1 and 2, and Liouville examined 3, finding, in the process, additional results related to 1 and 2. These investigations differed fundamentally from those described in the preceding section; for, when k, g, and l were not known, no workable explicit expression for the solution of (60) could be found. Thus all conclusions had to be derived from the equation itself, and these could only be of a qualitative nature. In this sense Sturm-Liouville theory anticipated Poincaré's approach to nonlinear differential equations.

Sturm and Liouville, in turn, had been anticipated by d'Alembert and Siméon Denis Poisson. D'Alembert [1769] had investigated the vibrating string of arbitrary varying thickness; this gave rise to equation (60), with $k = 1$ and $l = 0$, and with the boundary conditions $V(\alpha) = V(\beta) = 0$. He tried to establish the existence of one eigenvalue r_1, but his proof was insufficient. Still, it contained the important idea of comparing the equation with another having constant coefficients. D'Alembert's research does not seem to have influenced Sturm and Liouville. They did however, know and use Poisson's results. The latter had given a satisfactory, although laborious, proof that the eigenfunctions of (60)–(62) (for $l = 0$)

satisfy the orthogonality relations

$$\int_\alpha^\beta g(x)V_m(x)V_n(x)\,dx = 0 \qquad \text{for } m \neq n, \qquad (64)$$

from which he had correctly concluded that there are no complex eigenvalues [Poisson, 1826, 1835].

These were the only general results known about the eigenvalue problem (60)–(62) before the publication of Sturm's first paper [1836a] in which he investigated the more general equation,

$$\left(K(r,x)V_r'(x)\right)' + G(r,x)V_r(x) = 0 \qquad \text{for } x \in (\alpha, \beta), \quad (65)$$

subject to the boundary condition,

$$\frac{K(r,x)V_r'(x)}{V_r(x)} = h(r) \qquad \text{for } x = \alpha, \qquad (66)$$

where r is a real parameter. Sturm argued that for each value of r there exists a nontrivial solution, $V_r(x)$, to (65) and (66), and it is unique to within a multiplicative constant. The central part of his paper consisted of an investigation of the distribution of the roots of V_r. He first proved that under variation of r a root $x(r)$ of V_r can appear or disappear from the interval (α, β) only if it crosses one of the boundaries, $x(r) = \alpha$ or $x(r) = \beta$. For otherwise there must be a value of r for which V_r has a double root (see Figure 4), and according to (65) V_r would then be identically zero.

Hence it makes sense to follow a particular root x_r as r varies. To this end Sturm differentiated equation (65) with respect to r, denoting this process by δ to distinguish it from differentiation d

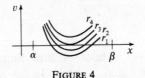

FIGURE 4

with respect to x; he obtained, after some manipulations (for convenience I write V instead of $V_r(x)$),

$$\delta V(KV') - V\delta(KV') = -V^2\delta\left(\frac{KV'}{V}\right)$$

$$= \left[-V^2\delta\left(\frac{KV'}{V}\right)\right]_{x=\alpha} + \int_\alpha^x V^2\delta G\, dx \quad (67)$$

$$- \int_\alpha^x (V')^2 \delta K\, dx.$$

Thus KV'/V is a decreasing function for all $x \in [\alpha, \beta]$ if the following conditions are fulfilled:

$$K > 0 \qquad \forall r, \forall x \in [\alpha, \beta],$$
$$G \text{ is an increasing function of } r, \qquad \forall x \in [\alpha, \beta],$$
$$K \text{ is a decreasing function of } r, \qquad \forall x \in [\alpha, \beta], \qquad (68)$$
$$\left[\frac{KV'}{V}\right]_{x=\alpha} = h(r) \text{ is a decreasing function of } r.$$

In fact, two of the three last quantities may be constant functions of r, so that the conditions are satisfied for the special case (60) and (61). That KV'/V decreases, means that it decreases from $+\infty$ to $-\infty$, where $V_r(x) = 0$, and then jumps again to $+\infty$.

Now, if x_r is a root of V_r, it is easy to see that

$$\frac{dr}{dx_r} = -\frac{\partial V}{\partial x}\bigg/\frac{\partial V}{\partial r}. \qquad (69)$$

From (67) Sturm further concluded that $\partial V/\partial r$ (or δV) and $\partial V/\partial x$ (or V') have the same sign when $V = 0$, and therefore

$$\frac{dr}{dx_r} < 0, \qquad (70)$$

which means that the roots x_r decrease as r increases.

This argument gives a taste of how Sturm drew qualitative conclusions from the equation itself. In his second paper of the same year [Sturm, 1836b], he applied this theorem to (60)–(61). If

r_n is an eigenvalue corresponding to (60)–(62), $V_{r_n} = V_n$ must satisfy (62), or

$$\frac{kV_n'(\beta)}{V_n(\beta)} = -H. \tag{71}$$

When r increases from r_n, $kV_r'(\beta)/V_r(\beta)$ decreases, and it can only be equal to $-H$ again after it has been $-\infty$, that is after $V_r(\beta)$ has been zero. Then a new root of $V_r(x)$ appears in (α, β), and since (61) prevents roots from leaving (α, β) through α, the next eigenfunction V_{n+1} has one more root in (α, β) than V_n. Moreover, Sturm proved that the roots of V_{n+1} separate those of V_n.

Using a similar procedure Sturm proved also what is now called his comparison theorem, and by comparing (60) with an equation with constant coefficients, he established that (60)–(62) has a (countable) infinity of real eigenvalues.

Sturm had already developed most of this remarkable theory in the period from 1829 to 1833, but it did not appear in full, before his friend Joseph Liouville invited him to publish it in the first volume of his journal.

As a young student Liouville himself had considered a problem similar to (46)–(48) [Liouville, 1830], and he had shown that it has only positive eigenvalues and that the eigenfunctions satisfy the orthogonality relations (69). This allowed him to calculate the "Fourier" coefficients A_n in (63) and to show that the Fourier series is convergent. Liouville's methods were strikingly different from Sturm's and were, in fact, not quite rigorous. Instead of Sturm's careful analysis of the oscillatory behavior of the eigenfunctions, based on an analysis of the equation itself, Liouville worked with a very awkward series expansion of the solution, drawing his conclusions from that. Although these conclusions were not quite rigorous the derivation of the series was still a milestone in the history of differential equations, for it was the first use of the method of successive approximations. Moreover, Liouville carefully proved the convergence of the series, in this way providing the first general, published existence theorem for solutions to differential equations.

In 1835–1836 Liouville improved his theory partly by building on Sturm's results and partly by refining his earlier arguments. He inserted the Fourier coefficients A_n, found by termwise application of the orthogonality relation (69), in (49) and obtained the Fourier series,

$$F(x) = \sum_{n=1}^{\infty} u_n(x) = \sum_{n=1}^{\infty} \frac{V_n(x) \int_{\alpha}^{\beta} g(x) V_n(x) f(x)\, dx}{\int_{\alpha}^{\beta} g(x) V_n^2(x)\, dx}. \quad (72)$$

His main goal [Liouville, 1837a] was to prove convergence of this series. By the so called Liouville transformation, he reduced the problem (60)–(72) to the simpler eigenvalue problem,

$$U''(z) + rU(z) = \lambda(z)U(z) \qquad \text{for } z \in (0, \gamma)$$

$$U'(z) - h_1 U(z) = 0 \qquad\qquad \text{for } z = 0 \qquad (73)$$

$$U'(z) + H_1 U(z) = 0 \qquad\qquad \text{for } z = \gamma,$$

and proved that the nth term u_n of the Fourier series (72) is numerically bounded by M/r_n, that is,

$$|u_n| \leqslant \frac{M}{r_n}. \quad (74)$$

He then rigorously established the asymptotic behavior of the eigenvalues,

$$\sqrt{r_n} \sim \frac{(n-1)\pi}{\gamma} + \frac{P_0}{(n-1)\pi}, \quad (75)$$

where P_0 is a constant. Combining this with (74), he obtained the estimate,

$$|u_n| < \frac{M'}{n^2}, \quad (76)$$

which implies the convergence of the Fourier series.

In order for his proof to work Liouville had assumed explicitly that g, k, f, and their first and second derivatives were finite and, implicitly, that f satisfied the boundary conditions. Later the same year [Liouville, 1837b], however, he devised a new convergence proof which relaxed these requirements. The central idea in the new proof was to use the asymptotic behavior of the eigenfunctions to show that the terms u_n of the Fourier series are of the form

$$u_n(x) = \frac{2}{\gamma}\cos nz \int_0^\gamma F(z)\cos nz\,dz + \frac{\psi(x,n)}{n^2}, \qquad (77)$$

where $\psi(x, n)$ is bounded. Thus the problem of convergence had been reduced to that of convergence of the ordinary trigonometric Fourier series, which had been established by Dirichlet [1829].

The convergence proof was Liouville's main achievement in this field. However, it did not follow from the proof that the limit of the Fourier series is the expanded function, i.e. that $F(x) = f(x)$ in (72). Liouville never succeeded in establishing this theorem rigorously, although he tried hard to do so; his very first paper on the subject [Liouville, 1836] was devoted to this question. Here he multiplied both sides of (72) by $g(x)V_m(x)dx$ and integrated from α to β. Using the orthogonality relations, he deduced that

$$\int_\alpha^\beta g(x)\varphi(x)V_m(x)\,dx = 0 \qquad \forall m = 1, 2, \ldots, \qquad (78)$$

where

$$\varphi(x) = F(x) - f(x). \qquad (79)$$

He then showed that for an arbitrary choice of j distinct points, $a_1, a_2, \ldots, a_j$ of (α, β), one can find a linear combination $P_j(x)$ of the first $j + 1$ eigenfunctions, such that P_j vanishes and changes sign at $a_1, a_2, \ldots, a_j$, and has no other roots. The P_j can be used to show that $\varphi(x)$ cannot change sign a finite number of times; for if $\varphi(x)$ changed sign at $a_1, a_2, \ldots, a_j$, one would have

$$\int_\alpha^\beta g(x)\varphi(x)P_j(x)\,dx \neq 0, \qquad (80)$$

which contradicts (78). From this observation Liouville concluded that $\varphi(x)$ must be zero, and whence that $F(x) = f(x)$.

However, this last inference is wrong, and Liouville soon realized it. In 1838, for example, he noted in his notebook [Liouville, Ms 3616(2), 56v] that φ may be different from zero in isolated points. Two years later he realized that there exist functions that oscillate infinitely often [Liouville, Ms 3616(5), 45–46] and that their existence clearly invalidates the whole argument. Moreover, in a note (probably from 1842) he wrote, after a similar argument, "In all this I only see the objection of $\mathscr{D}$. regarding the functions changing sign an infinite number of times" [Liouville, Ms 3627(2), 26v].[9]

Thus Liouville had been convinced of the insufficiency of his proof by a certain $\mathscr{D}$, probably his good friend Dirichlet, who had earlier taken great pains to exclude such infinitely often oscillating functions from his theorem on Fourier series. As a result of Dirichlet's objections Liouville made several unsuccessful attempts to repair the proof. In 1839 he designed a completely different proof and sent it to Dirichlet the following year [Tannery, 1910, 17–19].[10] However, this proof also contained a flaw, and Liouville finally gave up the search for a correct proof.

In addition to his work on second-order equations Liouville [1838] tried to generalize his and Sturm's theorems to certain higher-order equations; in 1838 he gave a course at the Collège de France on that subject. Thereafter his interest in the subject decreased, probably because even for third-order equations he had been unable to prove the theorem he considered the most important, namely that the Fourier series of a function f converges to f. There were good reasons for Liouville's failure, for the theorem generally fails for the third-order equations discussed by him.

In their papers, Sturm and Liouville gave a profound qualitative investigation of the questions raised by separation of variables in a general class of partial differential equations with variable coefficients. Of the "components" 1–6 (see Introduction) the only one

---

[9] My translation.

[10] Tannery's dating of the letter is wrong. See [Neuenschwander, 1984, 69].

they could not fully illuminate was 4. Sturm and Liouville succeeded in freeing their theory from those physical problems and partial differential equations from which it had emerged and were able to establish this theory as a self-contained subject.

Their work was so exhaustive that for the next forty years nothing happened in the field. Indeed, it was not until the late 1870s that there was renewed interest in Sturm-Liouville theory. We shall give only a brief sketch of these developments, which can be divided roughly into two mutually interacting categories— generalizations and rigorizations.

Among the generalizations are the discussions of higher-order equations published by Lord Rayleigh [1877] and Gustav Robert Kirchhoff [1879]. The singular Sturm-Liouville problems where, for example, $k(x)$ in (60) has zeros in the interval $[\alpha, \beta]$ were discussed by Ludwig Schläfli [1876] and later by Maxime Bôcher and others. The extension to the singular case made the general theory applicable to Bessel functions. Felix Klein [1881] also included polynomial solutions, for example, Lamé functions into the theory.

The most difficult part of the rigorization of Sturm-Liouville theory was to replace Liouville's essentially incorrect proof of the theorem that a Fourier series converges to the function that gives rise to it. Rigorous proofs for increasingly general functions were given by Henri Poincaré [1894], Vladimir Andreevich Steklov [1898], and Adolph Kneser [1904].

Poincaré's paper contained an idea which revolutionized the whole theory. He showed how to convert a Sturm-Liouville problem to an eigenvalue problem of an integral operator by way of its Green's function. This observation meant that the far reaching results on integral operators, obtained around 1900 by David Hilbert [1912] and others, had a direct impact on Sturm-Liouville theory. Thus the Hilbert spaces, created in this connection, shifted the interest from pointwise convergence of the Fourier series, to the easier question of convergence in L^2.

Moreover, Erhard Schmidt's geometric interpretation of Hilbert spaces [Schmidt, 1908] made it possible for mathematicians to imagine functions as vectors in a vector space and to interpret equation (78) as an orthogonality relation. The Fourier expansion of a function could be viewed geometrically as a decomposition of a vector with respect to an orthonormal system, and Liouville's great problem became a question of the completeness of this orthogonal system. This geometric reinterpretation was an enormous conceptual gain, without which it is hard to imagine the modern theory.

CONCLUDING REMARKS

From the time that partial differential equations first appeared on the scene in the middle of the 18th century, the method of separation of variables has been one of the main techniques used in their solution. In some cases the method has a simple physical interpretation; for example it provides a musical interpretation to vibrating bodies. This link to physics explains why essential parts of the method, such as eigenfunctions (simple modes) and their superposition, were understood even before the partial differential equations were introduced. When the method was first applied to partial differential equations, most mathematicians were of the opinion that it produced only special solutions, but after the work of Fourier, it was believed to be a general method (even in some cases where this is wrong—see Liouville's generalization).

Many problems were easily solved by the new method but, as with all good mathematical theories, it generated more new problems and techniques: for example, the study of curvilinear coordinates, which became a major element in differential geometry, and the eigenvalue problems for ordinary and later partial differential equations. The eigenvalue problems arising from the most important physical problems led to a variety of special functions, such as Bessel functions, spherical harmonics, Lamé functions, Mathieu functions etc., whose study occupied many mathematicians during the 19th century.

Problems of vibrations and heat conduction in materials with variable physical properties led, by separation of variables, to the more general eigenvalue problems of second-order ordinary differential equations. The investigations of these questions by Sturm and Liouville marked a turning point in the theory of differential equations. Earlier, the standard question in this field had been:

given a differential equation; find its solution as an analytic expression. Their theory taken together with contemporary existence theorems, broadened the question to: *given a differential equation; investigate some properties of the solution.*

Then almost half a century after their work appeared the Sturm-Liouville theory was revived and, together with the closely related theory of integral operators, it became the primary motivation for the creation of functional analysis at the beginning of this century (cf. [Dieudonné, 1981]). Thus the method of separation of variables, at first a useful method of solution, gave rise to some of the most important fields of mathematical analysis.

RECOMMENDED SECONDARY LITERATURE

Most of the topics discussed in this chapter are treated in more detail in the secondary literature. The standard work on the vibratory motions discussed in Section 1 and part of Section 2 is Truesdell's thorough and clear presentation [1960], but Cannon and Dostrovsky [1981] can also be recommended. The mathematical aspect of this theory is also covered in the extremely comprehensive work by Burkhardt [1908], which also includes much information on the development of special functions. The latter subject is also well treated by Bôcher in the historical Chapter IX of [Byerly, 1893]. Many historical notes on special functions can also be found in the articles of Wangerin and Burkhardt in the *Encyclopädie der Mathematischen Wissenschaften* (vol. II, 1, 2) and, in particular, in Watson's book on Bessel functions [Watson, 1922]. On the emergence of Sturm-Liouville theory the reader might find [Lützen, 1984a] illuminating. The implications for functional analysis are described in [Dieudonné, 1981] and [Birkhoff and Kreyszig, 1984]. Finally, Kline's book [1972] can be highly recommended as an introduction to all the subjects treated in this chapter as well as to related subjects.

REFERENCES

D'Alembert, J. R., 1743, *Traité de Dynamique*, David, Paris.

_____, 1746, "Recherches sur la courbe que forme une corde tenduë mise en vibration" and "Suite des Recherches..." *Histoire et Mémoires de l'Académie Royale des Sciences et Lettres de Berlin* 3 (1747), 214–249.

_____, 1750, "Addition au mémoire sur la courbe... " *Histoire et Mémoires de l'Académie des Sciences et Lettres de Berlin* 6 (1750), 355–360.

_____, 1769, "Extrait de différentes lettres de Mr. d'Alembert á Mr. de La Grange" *Histoire et Mémoires de l'Académie Royale des Sciences et Lettres de Berlin*, (1763 publ. 1770), 235–277.

Bernoulli, D., 1733, "Theoremata de oscillationibus corporum filo flexili connexorum et catenae verticaliter suspensae," *Commentarii Academiae Scientiarum Imperialis Petropolitanae* 6 (1732/33 publ. 1738), 108–122.

———, 1734, "Demonstrationes theorematum suorum de oscillationibus corporum filo flexili connexorum et catenae verticaliter suspensae," *Commentarii Academiae Scientiarum Imperialis Petropolitanae* 7 (1734-35) (publ. 1740), 162–173.

———, 1740, "Commentationes de oscillationibus compositis praesertim iis quae fiunt in corporibus ex filo flexili suspensis," *Commentarii Academiae Scientiarum Imperialis Petropolitanae* 12 (1740 publ. 1750).

———, 1742, "De sonis multifariis quos laminae elasticae diversimode edunt disquisitiones mechanico-geometricae..." *Commentarii Academiae Scientiarum Imperialis Petropolitanae* 13 (1741/43 publ. 1751), 167–196.

———, 1753, "Réfexions et éclaircissemens sur les nouvelles vibrations des cordes," *Histoire et Mémoires de l'Académie Royale des Sciences et Lettres de Berlin* 9, 147–172.

Bernoulli, J., 1727, "Theoremata selecta, pro conservatione virium vivarum demonstranda et experimentis confirmanda, excerpta ex epistolis datis ad filium Danielem" 11 Oct et 20 Dec (stil.nov) 1727, *Commentarii Academiae Scientiarum Petropolitanae* 2 (1727 publ. 1729), 200–207 = Opera Omnia 3, 124–130.

Birkhoff, G., and Kreyszig, E., 1984, "The Establishment of Functional Analysis," *Historia Mathematica* 11 (1984), 258–321.

Burkhardt, H., 1908, "Entwicklungen nach oscillirenden Funktionen und Integration der Differentialgleichungen der mathematischen Physik," *Jahresbereicht der Deutchen Mathematiker-Vereinigung X 2*, two volumes, Teubner, Leipzig.

Byerly, W. E., 1893, *An Elementary Treatise on Fourier's Series and Spherical, Cylindrical, and Ellipsoidal Harmonics*, Ginn and Company, Boston.

Cannon, J. T., and Dostrovsky, S., 1981, *The Evolution of Dynamics: Vibration Theory from 1687 to 1747*, Springer-Verlag, New York.

Clairaut, A. C., 1757, "Mémoire sur l'orbite apparente du soleil autour de la terre, en ayant égard aux perturbations produites par les actions de la Lune et des Planètes principales," *Histoire de l'Académie Royale des Sciences de Paris*, 1754, (pub. 1759), 52–564.

Dieudonné, J., 1981, *History of Functional Analysis*, North-Holland, Amsterdam.

Dirichlet, J. P. G. Lejeune, 1829, "Sur la convergence des séries trigonométriques qui servent à représenter une fonction arbitraire entre les limites données," *Journal für die Reine und Angewandte Mathematik* 4, 157–169. Werke I, 117–132.

———, 1837, "Sur les séries, dont le terme général dépend de deux angles, et qui servent à exprimer des fonctions arbitraires entre des limites données," *Journal für die Reine und Angewandte Mathematik* 17, 35–56 = Werke I, 273–306.

Engelsman, S. B., 1984, *Families of Curves and the Origin of Partial Differentiation*, North-Holland, Amsterdam.

Euler, L., 1743, "De integratione aequationum differentialium altiorum gradum," *Miscellanea Berolinensia* 7, 193–242 = Opera Omnia (1) XXII 108–149.

———, 1753, "Remarques sur les mémoires précédens de M. Bernoulli," *Histoire et Mémoires de l'Académie Royale des Sciences et Lettres de Berlin* 9, 1753 (publ. 1755), 196–222. = Opera Omnia (2) 10, 233–254.

———, 1759, "De motu vibratorio tympanorum," *Novi Commentarii Academiae Scientiarum Petropolitanae* 10 (1764 publ. 1766), 243–260 = Opera Omnia (2) 10, 344–359.

_____, 1774a, "De oscillationibus minimis funis libere suspensi," *Acta Academiae Scientiarum Petropolitanae* (1781:I publ. 1784), 157–177 = Opera Omnia (2) 11, 307–323.

_____, 1774b, "De perturbatione motus chordarum ab earum pondere oriunda," *Acta Academiae Scientiarum Petropolitanae* (1781 publ. 1784). = Opera Omnia (2) 11, 1, 324–334.

_____, 1777, "Disquisitio ulterior super seriebus secundum multipla cuiusdam anguli progredientibus," *Nova Acta Academia Scientiarium Petropolitanae* 11 (1793 publ. 1798), 114–132 = Opera Omnia (1) 16, 1, 333–355.

Fourier, J., 1822, *Théorie analytique de la chaleur*, Firmin Didot, Paris.

Hilbert, D., 1912, *Grundzüge einer algemeinen Théorie der linearen Integralgleichungen*, Teubner, Leipzig.

Kirchhoff, G., 1879, "Über die Transversalschwingungen eines Stabes von veränderlichem Querschnitt," *Monatsberichte der Berlin Akademie* (1879), 815–828 = *Annalen der Physik und Chemie* 10 (1880), 501–512.

Klein, F., 1881, "Über Körper, welche von confocalen Flächen Zweiten Grades begränzt sind," *Mathematische Annalen* 18, 410–427.

Kline, M., 1972, *Mathematical Thought from Ancient to Modern Times*, Oxford University Press, New York.

Kneser, A., 1904, "Untersuchungen über die Darstellung willkürlicher Funktionen in der mathematischen Physik," *Mathematische Annalen* 58, 81–147.

Lamé, G., 1833, "Mémoire sur les surfaces isothermes dans les corps solides en équilibre de température," *Annales de Chimie et Physique* (2) 53, 190–204.

_____, 1837, "Sur les surfaces isothermes dans les corps solides homogènes en équilibre de température," *Mémoires des savants étrangers de l'Académie Royale des Sciences* 5 = *Journal de Mathématiques Pures et Appliquées* 2, 147–183.

_____, 1839, "Mémoire sur l'équilibre des températures dans un ellipsoïde à trois axes inégaux," *Journal de Mathématiques Pures et Appliquées* 4, 126–163.

_____, 1859, "Leçons sur les coordonnées curvilignes et leurs diverses applications," Mallet-Bachelier, Paris.

Laplace, P. S., 1782, "Théorie des attractions des sphéroïdes et de la figure des planètes," *Mémoires de l'Académie Royale des Sciences*, Paris (1782 publ. 1785), 113–196.

Legendre, A. M., 1784, "Recherches sur la figure des planètes," *Mémoires des savants étrangers de l'Académie Royale des Sciences*, Paris, (1784 publ. 1787), 370–389.

Liouville, J., Ms, Notebooks in the Bibliothèque de l'Institut de France.

_____, 1830, "Mémoire sur la théorie analytique de la chaleur," *Annales de Mathématiques Pures et Appliquées* 21 (1830/31), 133–181.

_____, 1836, "Mémoire sur le développement des fonctions ou parties de fonctions en séries dont les divers termes sont assujettis à satisfaire à une même équation différentielle du second ordre, contenant un paramètre variable," *Journal de Mathématiques Pures et Appliquées* 1, 253–265.

_____, 1837a, "Second Mémoire sur le développement des fonctions..." *Journal de Mathématiques Pures et Appliquées* 2, 16–35.

_____, 1837b, "Troisième Mémoire sur le développement des fonctions..." *Journal de Mathématiques Pures et Appliquées* 2, 418–437.

_____, 1838, "Premier mémoire sur la théorie des équations différentielles linéaires, et sur le développement des fonctions en séries," *Journal de Mathématiques Pures et Appliquées* 3, 561–614.

_____, 1845, "Sur diverses questions d'analyse et de physique mathématique," *Journal de Mathématiques Pures et Appliquées* 10, 222–228.

_____, 1846, "Sur une transformation de l'équation $\sin\theta(d\sin\theta\frac{d\phi}{d\theta}/d\theta) + d^2\phi/d\omega^2 + n(n+1)\sin^2\theta \cdot \phi = 0$," *Journal de Mathématiques Pures et Appliquées* 11, 458–461.

Lützen, J., 1983, "Euler's Vision of a General Partial Differential Calculus for a Generalized Kind of Function," *Mathematics Magazine* 56, 299–306.

_____, 1984a, "Sturm and Liouville's Work on Ordinary Linear Differential Equations. The Emergence of Sturm-Liouville Theory," *Archive for History of Exact Sciences* 29, 309–376.

_____, 1984b, "Joseph Liouville's Work on the Figures of Equilibrium of a Rotating Mass of Fluid," *Archive for History of Exact Sciences* 30, 113–166.

Mathieu, E. L., 1868, "Mémoire sur le mouvement vibratoire d'une menbrane de forme élliptique," *Journal de Mathématiques Pures et Appliquées* (2) 13, 137–203.

Neuenschwander, E., 1984, "Joseph Liouville (1809–1882): Correspondance inédite et documents biographiques provenant de différentes archives parisiennes," *Bulletino di Storia delle Scienze Matematice* 4, 55–132.

Poincaré, H., 1894, "Sur les Équations de la Physique Mathématique," *Rendiconti del Circolo Matematico di Palermo* 8, 57–155.

Poisson, S. D., 1826, "Note sur les racines des équations transcendantes," *Bulletin de la Societé Philomatiques* (1826), 145–148.

_____, 1835, *Théorie mathématique de la chaleur*, Bachelier, Paris.

Ravets, J. R., 1961, "Vibrating strings and arbitrary functions. Logic of personal knowledge," *Essays presented to M. Polanyi on his 70th birthday*, London.

Lord Rayleigh, 1877, *The Theory of Sound*, London.

Robartes, F., 1692, "A discourse concerning the musical notes of the trumpet, and the trumpet marine, and of defects of the same," *Philosophical Transactions* 17, 559–563.

Schläfli, L., 1876, "Über die Convergenz der Entwicklung einer arbiträren Funktion $f(x)$ nach den Bessel'schen Funktionen....," *Mathematische Annalen* 10, 137–142.

Schmidt, E., 1908, "Über die Auflösung linearer Gleichungen mit unendlich vielen Unbekannten," *Rendiconti del Circolo Matematico di Palermo* 25, 53–77.

Steklov, W., 1898, "Sur le problème de refroidissement d'une barre hétérogène," *Comptes Rendus de l'Académie des Sciences de Paris* 126, 215–218.

Sturm, C., 1836a, "Mémoire sur les équations différentielles linéaires du second ordre," *Journal de Mathématiques Pures et Appliquées* 1, 106–186.

_____, 1836b, "Mémoire sur une classe d'équations à différences partielles," *Journal de Mathématiques Pures et Appliquées* 1, 373–444.

Tannery, J., 1910, *Correspondence entre Lejeune Dirichlet et Liouville*, Gauthier-Villars, Paris.

Taylor, B., 1713, "De motu nervi tensi," *Philosophical Transactions of the Royal Society London* 28, 26–32.

———, 1715, *Methodus incrementorum directa et inversa*, London, 1715.

Truesdell, C., 1960, *The Rational Mechanics of Flexible or Elastic Bodies 1638–1788, Leonardi Euleri Opera Omnia* (2) XI 2. Orell Füssli Turici.

Wallis, J., 1677, "On the trembling of consonant strings, a new musical discovery," *Philosophical Transactions* 12, 839–842.

Watson, G. N., 1922, *A Treatise on the Theory of Bessel Functions*, Cambridge University Press, New York, 2nd. ed., 1944.

Weber, H., 1869, "Ueber die Integration der partiellen Differentialgleichung $(\partial^2 u/\partial x^2) + (\partial^2 u/\partial y^2) + k^2 u = 0$," *Mathematische Annalen* 1, 1–36.

Youschkevich, A. P., 1976, "The concept of function up to the middle of the 19th century," *Archive for History of Exact Sciences* 16, 37–85.

THE BEGINNINGS OF ITALIAN ALGEBRAIC GEOMETRY

Jean Dieudonné

In the history of algebraic geometry in the nineteenth century, one may discern several more or less overlapping sequences:

1. 1825-1890. The projective theory of algebraic curves and surfaces was developed first for curves and surfaces embedded in $P_2(\mathbf{C})$ and $P_3(\mathbf{C})$ and later (after 1870) in a projective space, $P_r(\mathbf{C})$, of arbitrary dimension. This consisted, on the one hand, in studies of particular curves and surfaces of low degree and, on the other hand, in what one may call "Plückerian" formulas (i.e., based on the pattern of the famous "Plücker formulas" for plane curves) involving more and more numerous projective invariants and more and more complicated singularities. Beginning with G. Salmon, A. Cayley, and G. Halphen, almost all algebraic geometers of the nineteenth century contributed more or less to that trend ([Baker, 1933] contains many of their results).

2. 1857-1900. During this period the birational geometry of algebraic curves, in a projective space of arbitrary dimension, was created.

3. 1860-1890. Cremona Transformations in the projective plane and rational surfaces in projective spaces of arbitrary dimension were studied.

4. 1870-1890. The earliest investigations of the birational geometry of algebraic surfaces took place.

5. 1890-1900. The work of G. Castelnuovo and F. Enriques on linear systems of curves and surfaces appeared.

The choice of 1900 as a terminal date is not at all arbitrary. By the end of this paper, it should be clear why this date can be considered as the beginning of a new period in algebraic geometry.

BIRATIONAL CORRESPONDENCE AND DIVISORS

Until 1900 algebraic geometry was dominated by two fundamental notions: birational correspondences and divisors (or, equivalently, *linear systems*). Already in the years 1820–1860, nonlinear birational transformations had been considered in projective geometry, but without any attempt at a general theory. Some were defined in the complement of a finite subset of the whole projective plane, such as the *quadratic transformation*

$$(x, y, z) \mapsto (yz, zx, xy), \tag{1}$$

which is defined in the complement of the vertices of the triangle (union of the lines $x = 0$, $y = 0$, $z = 0$) and injective in the complement of the triangle. Other transformations were defined only on an algebraic curve Γ: for example, to each point M on Γ is associated the point of contact of the tangent corresponding to M for the curve Γ' transformed from Γ by duality.

However, it is only with B. Riemann that birational correspondences reached a central position. It is well known that the chief objects of his study were not curves, but algebraic *functions* and their integrals, or equivalently, meromorphic functions on their Riemann surfaces; all algebraic curves having the same meromorphic functions on their Riemann surfaces were, therefore, equivalent for him, and this meant that they could be deduced from one another by birational transformations.

The concept of *divisor* also occurred (in a particular case and without a name) in Riemann's fundamental paper of 1857: having given a topological definition of the *genus p* of a compact Riemann surface S, he examined the possibility of finding on S a nonconstant meromorphic function f having at most poles of the first order at m given points $a_1, a_2, \ldots, a_m$ of S. If $v_x(f)$ is the order (positive or negative) of f at a point $x \in S$, this condition means that

$$v_{a_j}(f) \geqslant -1 \qquad \text{for } 1 \leqslant j \leqslant m,$$

$$v_x(f) \geqslant 0 \qquad \text{for all other points of } S. \qquad (2)$$

If V is the vector space over $\mathbf{C}$, contained in the field K of meromorphic functions on S, defined by the inequalities (2), then the problem is to evaluate dim V and to determine if dim $V \geqslant 2$.

This problem may be generalized in a way that is substantially equivalent to the ideas used in 1882 by R. Dedekind and H. Weber to develop the theory of algebraic curves: a *divisor* on S is a family $D = (\alpha_x)_{x \in S}$ of integers (of any sign) which are 0 except at a finite number of points $x \in S$. The set $L(D)$ of all $f \in K$, such that $v_x(f) \geqslant -\alpha_x$ for all $x \in S$, is then a finite-dimensional vector space over $\mathbf{C}$, and the general problem is to determine $l(D) = \dim L(D)$. If one writes $\deg(D) = \Sigma_{x \in S} \alpha_x$, the methods used by Riemann show that

$$l(D) \geqslant \deg(D) + 1 - p. \qquad (3)$$

When addition is defined componentwise, the divisors form an abelian group $\mathscr{D} (= \mathbf{Z}^{(S)})$. For each $f \in K$ (not equal to 0), one notes $(f)_0$ the sum of the $v_x(f)$ which are greater than 0 (i.e., zeroes of f) and $(f)_\infty = (1/f)_0$ (the poles of f); the divisors $(f) = (f)_0 - (f)_\infty$, are called the *principal divisors*, and form a subgroup $\mathscr{P}$ of $\mathscr{D}$; one proves that $\deg(f) = 0$. The condition $f \in L(D)$, can be written as

$$(f) + D \geqslant 0 \qquad (4)$$

(positive divisors (α_x) being defined as those for which $\alpha_x \geqslant 0$ for all x in S). If $D' - D \in \mathscr{P}$, then $\deg(D') = \deg(D)$ and $l(D') = l(D)$, and one says that D and D' are *linearly equivalent* divisors. Furthermore, the relation $(f) + D = (g) + D$ implies that f/g is a nonzero constant; the set $|D|$ of *positive* (or *effective*) divisors equivalent to D is therefore the projective space derived from $L(D)$, having dimension $l(D) - 1$.

The group structure on $\mathscr{D}$ enabled Dedekind and Weber to express the difference of the two sides of (3) as $l(\Delta - D)$, so that

$$l(D) - l(\Delta - D) = \deg(D) + 1 - p, \qquad (5)$$

where Δ is a divisor in a well determined class, the so-called *canonical class*, for which

$$\deg(\Delta) = 2p - 2, \qquad l(\Delta) = p - 1. \qquad (6)$$

This result was essentially proved first by Riemann's student G. Roch, and is called the *Theorem of Riemann-Roch* for curves; Roch used the language of abelian integrals, while Dedekind and Weber defined Δ by means of differential forms on S.

However, mathematicians of the German school of R. Clebsch and M. Noether, as well as geometers of the Italian school who followed them, defined the canonical class by means of another interpretation of divisors, the *linear series* on curves.

LINEAR SERIES AND THE BRILL-NOETHER THEORY

If Γ is a smooth algebraic curve in $P_2(\mathbf{C})$, it may be identified with its Riemann surface. Let E be a vector subspace of a vector space $L(D)$ for a divisor D on Γ; then if $\dim(E) = r + 1 \geqslant 0$, the rational functions $f \in E$ may be written as

$$\sum_{j=0}^{r} \lambda_j \frac{P_j(x, y, z)}{Q(x, y, z)}, \qquad (7)$$

where P_j and Q are homogeneous polynomials of the same degree without nonconstant common factor; the λ_j are complex numbers at least one of which is not 0; and it is assumed that the numerator never vanishes identically on Γ. It follows, then, that

$$(f) + D = \left(\sum_{j=0}^{r} \lambda_j P_j \right)_0 - (Q)_0 + D; \tag{8}$$

in Riemann's original problem, if $(f) + D \geqslant 0$, the divisor (8) is just $(\sum_j \lambda_j P_j)_0$, that is, the set of points of intersection (counted with their multiplicities) of Γ with the curve of equation

$$\sum_{j=0}^{r} \lambda_j P_j(x, y, z) = 0; \tag{9}$$

these sets of points constitute the *linear series* associated with E, usually written g_m^r if m is the number of points in each set. Conversely, for such a series (9), one may write

$$\left(\sum_{j=0}^{r} \lambda_j P_j \right)_0 = \left(\sum_{j=0}^{r} \lambda_j^0 P_j \right)_0 + (f), \tag{10}$$

where the λ_j^0 are particular values of the λ_j, $f = (\sum_j \lambda_j P_j)/(\sum_j \lambda_j^0 P_j)$, and these positive divisors form a linear projective variety in $|D|$, where $D = (\sum_{j=0}^{r} \lambda_j P_j)_0$.

A linear series consists of sets of points defined by equation (9) when the λ_j vary arbitrarily; they have the same cardinal m (when multiplicity of intersection is taken into account); the linear series is said to be *complete* if the vector space $E = L(D)$ for the corresponding divisor D. A linear series may have *fixed points* (i.e., points belonging to all the sets of the series, with some multiplicity).

The concept of linear series on a curve (introduced without mention of divisors) was the starting point for A. Brill and M. Noether [1873] in their fundamental paper on algebraic curves. Following Abel, Riemann had shown that if a curve Γ of nonhomogeneous equation $F(s, z) = 0$ has at most double points, then the abelian integrals of the first kind may be written as

$$\int \frac{Q(s, z) \, dz}{F_s'} \tag{11}$$

where $Q(s, z) = 0$ is the equation of a curve passing through the double points of Γ (an *adjoint* of Γ). R. Clebsch and P. Gordan expressed this in projective coordinates, showing that Riemann's formula for the genus of a curve Γ of degree n having d double points could be written as

$$p = (1/2)(n-1)(n-2) - d. \tag{12}$$

More generally, Brill and Noether considered curves with only *ordinary* multiple points, i.e., having k distinct tangents if the multiplicity is k. The genus of such a curve of degree n is then given by

$$p = (1/2)(n-1)(n-2) - (1/2)\sum_j k_j(k_j - 1), \tag{13}$$

with summation on the multiple points. Algebraic curves (of any degree) having a multiple point of order at least $k_j - 1$ at each point of Γ of order k_j were called *adjoints* of Γ. Their method, which will not be described here in detail, rests on the famous theorem of Noether that provides a geometric description of the ideal $\{AP + BQ\}$, generated by the two polynomials P and Q; their main results (including some further ones of Noether) follow:

All *complete* linear series g_m^r can be obtained by intersecting Γ with the curves of the set Φ_k of *all* adjoints of degree k, or by the curves of Φ_k passing through fixed points on Γ.

For $k = n - 3 + \alpha$, where $\alpha > 0$, $m = 2p - 2 + n\alpha$ and $r \geqslant p - 2 + n\alpha$ for the series cut by Φ_k. For $k = n - 3 - \alpha$, where $\alpha \geqslant 0$, $m = 2p - 2 - n\alpha$ and $r = p - 1 - n + \alpha(\alpha + 3)/2$ for the series cut by Φ_k.

The series cut by Φ_{n-3} is called the *canonical series* g_{2p-2}^{p-1}; it is the *only* linear series g_m^r with $m = 2p - 2$ and $r = p - 1$, and it has no fixed points. Any such series $g_m^r \subset g_{2p-2}^{p-1}$ is said to be *special*; in this case $m - r \leqslant p - 1$, and the converse is true if the series g_m^r is complete. When g_m^r is *complete* and *nonspecial*, $m - r = p$.

The Riemann-Roch theorem takes on the following form: suppose g_m^r is *complete* and *special*; then for any set $G_m \in g_m^r$, the curves of Φ_{n-3} passing through the points of G_m (with their multiplicity) form a projective space of dimension

$$r' = p - 1 - m + r \quad (\textit{index of speciality of } g_m^r). \tag{14}$$

These results for curves having only ordinary singularities gave, in fact, the first complete proofs of Riemann's results (since his transcendental proofs depended on many undefined topological notions and on the controversial "Dirichlet Principle"). This is due to the fact that *any* algebraic plane curve with *arbitrary* singularities may be transformed birationally into a curve with only ordinary multiple points. That result was proved by Noether [Brill and Noether, 1873] as follows: when the multiple points of Γ are not all ordinary, the expression (13) is greater than the genus; Noether showed that by a suitably chosen quadratic transformation, Γ is transformed into a curve for which the expression (13) has *decreased*; this process may be repeated until one reaches a curve for which all multiple points are ordinary.

CREMONA TRANSFORMATIONS AND RATIONAL SURFACES

Italian algebraic geometry got started in the early 1860s with the papers of L. Cremona (1830–1903), who investigated the structure of the *most general* birational transformations of the projective plane $P_2(\mathbf{C})$ onto itself. Such a transformation must be defined by

$$T:(x, y, z) \mapsto (P_1(x, y, z),\ P_2(x, y, z),\ P_3(x, y, z)), \quad (15)$$

where the P_j $(1 \leqslant j \leqslant 3)$ are three homogeneous polynomials having the same positive degree n and no nonconstant common factors; T is not defined at the *base points* where the polynomials P_j all vanish. The inverse T^{-1} transforms the set of lines $\lambda_1 x' + \lambda_2 y' + \lambda_3 z' = 0$ into a *net* of curves in the plane,

$$\lambda_1 P_1(x, y, z) + \lambda_2 P_2(x, y, z) + \lambda_3 P_3(x, y, z) = 0, \quad (16)$$

any two of which have only *one* common point outside the base points (the net is then called *homaloidal*). The conditions imply that the curves (16) are *rational* and satisfy the relations

$$(n-1)(n-2) - \sum_i k_i(k_i - 1) = 0 \quad (17)$$

and

$$n^2 - \sum_i k_i^2 = 1, \tag{18}$$

where the summations are over the base points and the k_i are their multiplicities. For small values of n, this led to the determination of all the corresponding Cremona transformations by several mathematicians, including M. Noether and E. Bertini (1846–1933).

A similar geometric interpretation of *general linear systems* of plane curves

$$\sum_{j=0}^{r} \lambda_j P_j(x, y, z) = 0 \tag{19}$$

(where the P_j are linearly independent homogeneous polynomials of the same degree n, without nonconstant common factor) seems to have first been proposed in 1870 by Brill. Base points having been defined as above, the mapping

$$T:(x, y, z) \mapsto \left(P_j(x, y, z)\right)_{0 \leqslant j \leqslant r} \tag{20}$$

is defined in the complement Ω of the finite set of base points and takes its values in $P_r(\mathbf{C})$; the closure S of $T(\Omega)$ is an algebraic surface, which was proved in 1894 by Castelnuovo [1896] to be *rational* (i.e., birationally equivalent to the projective plane); the complement of $T(\Omega)$ in S consists of rational algebraic curves, each "blown up" of one of the base points O_k, and consisting of the limits of points $T(M)$ when M tends to O_k along all directions. This idea was actively developed after 1870 by a second generation of Italian geometers, among whom the most prominent were G. Veronese (1857–1917), P. Del Pezzo (1859–1936), and C. Segre (1863–1924). They obtained many beautiful geometric results on particular rational surfaces: for instance, if one takes for (20) the family of all conics in the plane, one obtains the Veronese surface in $P_5(\mathbf{C})$, a surface of degree 4 on which all irreducible algebraic curves have even degree. If one takes for (20) all cubics passing through $9 - m$ base points with $3 \leqslant m \leqslant 9$, one gets the Del Pezzo surfaces S_m, which have degree m and are not contained in a hyperplane of $P_m(\mathbf{C})$. If $m = 3$ and the six base points are not on a conic, S_3 is the most general cubic surface; the six base points are "blown up" in six straight lines on S_3, and there are on S_3 twenty-one other lines which are transforms by T of the fifteen lines joining two base points and of the six conics passing by five of them.

Another application yields *rational ruled surfaces*: the system (20) then consists of all curves of degree n with n base points, one of which is multiple of order $n - 1$, and the remaining $n - 1$ others are simple; the surface is then a ruled surface of degree n in $P_{n+1}(\mathbf{C})$, and the lines generating S are the transforms of the lines through the multiple base point.

Finally, one can take for (20) the system of *all adjoints* of a high enough degree of a plane curve Γ with ordinary singularities; then $T(\Gamma)$ is a *smooth* curve in some $P_r(\mathbf{C})$ with $r \geqslant 3$, and by a suitable projection on a 3-dimensional subspace, one gets a smooth curve in $P_3(\mathbf{C})$ birationally equivalent to Γ. It was shown by C. Segre [1887, 1889] that any two such "models" of Γ are *projectively* equivalent.

THE BEGINNINGS OF THE THEORY OF ALGEBRAIC SURFACES

Even before the publication of the Brill-Noether theory of algebraic curves, there had been attempts to study algebraic surfaces in a similar way. The first idea was to generalize Riemann's methods based on abelian integrals and adjoint curves: double integrals $\iint R(x, y, z) \, dx \, dy$ were considered on a surface S defined by a nonhomogeneous equation $f(x, y, z) = 0$ of degree n. At first attention was concentrated on surfaces having as singularities at most a double curve L with at most triple points (also triple for the surface) and a finite number of "pinch points" where the two tangent planes coincide; these were, again, considered to be "*ordinary*" singularities. It is then not difficult to see that the double integrals, which are finite everywhere, can be written as

$$\iint \frac{Q(x, y, z) \, dx \, dy}{f_z'} \tag{21}$$

where $Q(x, y, z) = 0$ is a surface of degree $n - 4$ containing the double curve, again called an *adjoint* surface to S (see, for example, [Picard and Simart, 1897–1903]). This apparently led Clebsch [1868], in a short note appearing in the *Comptes Rendus* in which he did not mention integrals at all, to call the *genus* of S the maximum number of linearly independent polynomials Q defining adjoints of degree $n - 4$. In 1870 M. Noether showed, by sheer computation, that (for surfaces with ordinary singularities) the integrand in (21) is transformed into a similar one by a birational transformation of S into another surface with ordinary singularities (he even did this for hypersurfaces in a projective space of arbitrary dimension) [Noether, 1870]. A little later H. Zeuthen [1871] and Noether [1875] endeavored to compute Clebsch's "genus" by a formula which, for curves, would correspond to (13): "Plückerian" type formulas showed that for a general curve L in $P_3(\mathbf{C})$ of degree d, genus π, and having t triple points, the number of conditions for a surface of degree m to contain L is

$$md - 2t - \pi + 1. \tag{22}$$

Therefore, Clebsch's definition of a surface of degree n having L as a double curve would give, for the value of the "genus,"

$$(n - 1)(n - 2)(n - 3)/6 - (n - 4)d + 2t + \pi - 1. \tag{23}$$

In fact, Zeuthen and Noether showed that this number is invariant under birational transformations of surfaces having ordinary singularities. However, if S is a *ruled surface* of degree n, there are in general *no* surfaces of degree $n - 3$ (let alone $n - 4$) containing its double curve; so, the Clebsch genus should be equal to 0. But, in 1871 Cayley observed that if the plane sections of S have genus p, the value of (23) is $-p$. Thus it appeared very early that, for every surface with ordinary singularities, there were *two* invariantly attached numbers—the number p_g, defined by Clebsch and called the *geometric genus*, and the number (23), called the *arithmetic genus* and now written p_a; in all cases, $p_a \leqslant p_g$.

Other invariants under birational transformation—such as the "Curvengeschlecht" of Noether and what was later called the

"Zeuthen-Segre Invariant"—obtained by similar "Plückerian" methods during the early years of the study of algebraic surfaces will not be discussed here. The emphasis, here, on ordinary singularities may appear strange at first sight; but it stems from the conjecture that *any* algebraic surface is birationally equivalent to a *smooth* surface in $P_5(\mathbf{C})$. Such a surface, by a suitable projection on $P_3(\mathbf{C})$, is transformed into a surface with ordinary singularities. Many mathematicians endeavored to prove this conjecture, but it was only in 1935 that a completely correct proof was obtained by the American mathematician R. Walker [1935].

To give a complete picture of the early history of the theory of algebraic surfaces, one should also mention the special results concerning ruled surfaces, chiefly those of C. Segre [1887, 1889]; these are described in [Baker, 1933].

The remainder of this paper is devoted to the new life in algebraic geometry brought by the arrival on the scene of two younger scholars: G. Castelnuovo (1864–1952), followed a few years later by F. Enriques (1871–1946), and their fruitful collaboration.

THE LINEAR SYSTEMS OF CURVES ON AN ALGEBRAIC SURFACE

Their avowed purpose, as explained in Enriques' expository paper [1896], was to build up an *intrinsic* theory of algebraic surfaces, completely independent of any embedding in a projective space and dealing with properties invariant under birational transformations. That goal could not be completely reached before the existence of smooth models was proved and sheaf cohomology allowed one to work exclusively on such models (see, for instance, [Beauville, 1978] and [Mumford, 1966]). However, it was Castelnuovo and Enriques who took the first steps in that direction, by building their work around the central concept of a *linear system* of curves, the substitute in their theory of what was to become later (on a smooth surface) the concepts of divisor and of an invertible sheaf.

Of course, the definition of a divisor on a curve can be extended immediately to algebraic varieties of arbitrary dimension m by replacing points with irreducible subvarieties of dimension $m - 1$, and by using the fact that the set of zeroes of any polynomial on the variety has a closure which is the union of a finite number of such subvarieties (with suitable multiplicities). For a surface embedded in a projective space $P_r(\mathbf{C})$, the relation between positive divisors and the family of curves of intersections of the surface with a linear family $\sum_{j=0}^r \lambda_j P_j = 0$ of hypersurfaces is then the same as the one described above for curves, and one may define in a similar way the *dimension* $l(D)$, the *complete* linear systems $|D|$, and the *sum* and *difference* of linear systems. Castelnuovo and Enriques did not follow this path, choosing instead the pattern of the Brill-Noether theory, which led to some difficulties in defining the preceding notions (see Chapter II of [Zariski, 1934]).

The first paper [Castelnuovo, 1892] dealing with linear systems was a thorough study of the linear systems of *plane* curves; the main tool was the Brill-Noether theory of linear series on a curve (an idea that had been advocated a few years earlier by C. Segre). A linear system was defined by Castelnuovo to consist of curves of given degree n, having multiple points of order at least ν_j at given points A_j of the plane. Along with the *effective* dimension ρ of the system (the number of curves of the system passing through a generic point of the plane), Castelnuovo introduced the *virtual* dimension ρ', given by

$$\rho' = (1/2)\left\{ n(n+3) - \sum_j \nu_j(\nu_j + 1) \right\}, \qquad (24)$$

which would be equal to ρ if the conditions imposed on the curves at each A_j were independent (an example in which $\rho \neq \rho'$ is given by $n = 3$, all $\nu_j = 1$, and the A_j are the intersections of two cubics). The difference $s = \rho - \rho'$ is the *superabundance* of the system, which is said to be *regular* if $s = 0$. With the system $|C|$ Castelnuovo associated its *adjoint* system $|C_a|$, which consists of curves of degree $n - 3$ having multiple points whose orders are at least $\nu_j - 1$ at each A_j; he defined the *virtual genus* of $|C|$ to be the number

$$p' = (1/2)\left\{ (n-1)(n-2) - \sum_j \nu_j(\nu_j - 1) \right\} \qquad (25)$$

(one more than the virtual dimension of $|C_a|$). Finally, the *degree* of $|C|$ was defined by

$$d = n^2 - \sum_j \nu_j^2 \qquad (26)$$

(which, in the simplest cases, is the number of intersections—other than A_j—of two generic curves of $|C|$). Castelnuovo derived the relation

$$d = \rho' + p' - 1 \qquad (27)$$

and proved that the numbers ρ' and p' are *invariant* under Cremona transformations of the plane. He emphasized the case of systems $|C|$ for which the generic curve is *irreducible* (by an extension of a theorem of Bertini [1882], this always occurs except for relatively unimportant linear systems); these are called *irreducible* systems, and the virtual genus p' is then the effective genus p of that generic curve; the *characteristic series* $g_d^{\rho-1}$ cut by $|C|$ on one of its generic curves is then *complete*, from which the inequality

$$d - \rho + 1 \leqslant p \qquad (28)$$

follows, equality holding only if $s = 0$ (or, equivalently, if the characteristic series is nonspecial). A deeper study of the *pure* adjoint system (consisting of the variable curves of $|C_a|$) led Castelnuovo finally to the remarkable result,

$$\rho' \leqslant 2p - p_a' + 8, \qquad (29)$$

where p_a' is the virtual genus of the pure adjoint system, from which he was able to deduce many results on the relations between the invariants of a linear system; for example, if $\rho > 3p + 5$, all the curves of $|C|$ are rational.

Two years later Enriques published the first paper [1894] on linear systems of curves on an *arbitrary* algebraic surface; it was further developed and systematized in a long expository paper [Enriques, 1896]. His central problem was to define what should replace, for surfaces, the *canonical series* on a curve. If he had followed the method of Dedekind and Weber, he should have considered (at least on a smooth surface) the divisor associated with meromorphic 2-forms on the surface. When $p_g > 0$, the corresponding linear system would be cut on the surface by the adjoint surfaces of degree $n - 4$ (if n is the degree of the surface), and it would be the canonical system one looks for. But if $p_g = 0$, then the divisor of 2-forms is *negative*; it was not until 1905 that Italian geometers used this notion (under the name of "virtual curve"). So, Enriques introduced, in a roundabout way, an equivalent of the canonical divisor: when $p_g > 0$ and $|K|$ is the canonical system, he called $|K + D|$ the *adjoint* of $|D|$, written $|D'|$; for any two linear systems $|D_1|$ and $|D_2|$, one has then

$$|D_1' + D_2| = |D_1 + D_2'| = |(D_1 + D_2)'| \tag{30}$$

(for simplicity we have surpressed some additional conditions that must be made about the base points of the systems under consideration). What Enriques proved is that, even if $p_g = 0$, it may be possible to define, for any irreducible linear system, an "adjoint system" $|D'|$ having the property (30). This is the case, for example, if adjoint surfaces of degree $n - 3$ exist (again, for a surface with ordinary singularities); for, if $|C|$ is the linear system of all plane sections of the surface, then $|C'|$ is the system cut by these adjoint surfaces. On the other hand, for *any* linear system $|D|$ of dimension $r \geqslant 2$, there is a birational transformation of the surface that sends $|D|$ into a linear system containing the plane sections of the transformed surface; one has only to use Brill's method (as described above) for linear systems of plane curves: if $|D|$ is defined to be $\sum_{j=0}^{r} \lambda_j P_j = 0$, with $r \geqslant 2$, one considers the birational transformation

$$T : (x, y, z, t) \mapsto \left(P_j(x, y, z, t) \right)_{0 \leqslant r \leqslant 2}.$$

(Later, other methods which do not use any projective embeddings were given to define the adjoint system $|C'|$; see [Zariski, 1934, p. 65].) This use of plane sections was considered in a more general context by Castelnuovo in two papers, published simultaneously with the preceding one, *on the theme of Hilbert's polynomial* (which, for large enough values of m, gives the dimension of the vector space of homogeneous polynomials of degree m belonging to a given ideal). Castelnuovo [1892] first completed the earlier result by the following lemma. Given a system of base points with finite multiplicities v_j, let $|C^n|$ be the linear system of plane curves of degree n defined by these conditions: suppose that $|C^n|$ and $|C^{n+1}|$ exist and are both regular and complete; then $|C^{n+2}|$ is also complete (this is seen by considering the two systems of curves of degree $n + 2$ obtained by adjoining to $|C^{n+1}|$ two fixed lines). Now consider, in $P_3(\mathbf{C})$, a set of base curves with given multiplicities, and for each n let $|F_n|$ be the linear system of surfaces of degree n having these base curves; assuming it has no fixed components, let Σ_ω^n be the linear system cut by the surfaces of $|F_n|$ on a generic plane ω; in general, Σ_ω^n is neither complete nor regular; its dimension is therefore

$$\{(n+1)(n+2)/2\} - 1 - k + s_n - \delta_n, \qquad (31)$$

where s_n is the superabundance of Σ_ω^n, δ_n its defect, and k is a constant.

For each n such that $|F_n|$ exists, let $|\Gamma_\omega^n|$ be the *complete* linear system of curves of degree n on the generic plane ω having the same base points as Σ_ω^n. By considering, in the space $P_3(\mathbf{C})$, a pencil of planes ω passing through a generic line A, Castelnuovo showed that for any plane α containing A and any curve in $|\Gamma_\alpha^n|$, it is possible to find a $k \geqslant 0$ such that there is a surface in $|F_{n+k}|$ which cuts on α the curve decomposed in the given one and A taken k times. Using the lemma on linear systems of plane curves, he was able to show that for large enough n the system $|F_n|$ cuts on *any* generic plane ω a *complete* and *regular* system. Therefore, if the number n in (31) is permitted to grow, starting with the smallest value for which $|F_n|$ exists, there is a first value i such that $|\Gamma_\omega^n|$ is regular for all $n > i$ and, further, a value $l \geqslant i$ such that for $n > l$, Σ_ω^n is both complete and regular. This implies that for $n \geqslant l - 1$, the dimension of $|F_n|$ is given by

$$r_n' = \binom{n+3}{n} - kn + k' - 1, \qquad (32)$$

where k' is a constant; r_n is called the *virtual dimension* of $|F_n|$ (for all n). For $i - 1 \leqslant n < l - 1$, the effective dimension of $|F_n|$ is

$$r_n = r_n' + \sum_{j=n+1}^{l-1} \delta_j$$

and for $n < i - 1$

$$r_n = r_n' + \sum_{j=n+1}^{l-1} \delta_j - \sum_{j=n+1}^{i-1} s_j.$$

These results may now be applied to the case in which, for a surface S of degree n with ordinary singularities, $|F_m|$ is the system of *adjoint surfaces* of S of degree m; for $m \geqslant n - 1$, these surfaces always exist, and the curves they cut on a generic plane are the adjoints of degree m of the intersection of the plane with S; hence they always form a *regular system*. Furthermore, $r_{n-4} = p_g - 1$ and $r_{n-4}' = p_a - 1$, and it follows, therefore, that one has the fundamental identity (first proved by Enriques [1896]),

$$p_g - p_a = \sum_{h \geqslant n-3} \omega_h, \tag{33}$$

where ω_h is the defect of the system of curves cut on a generic plane by the adjoints of S of degree $h \geqslant n - 4$. The difference $q = p_g - p_a$ is called the *irregularity* of S, a *regular* surface being such that $p_g = p_a$.

In addition, Enriques showed that this result can be put in a form independent of the embedding of S in projective space. For *any* irreducible linear system $|D|$ of curves on S, the series cut on a generic curve D of $|D|$ by the adjoint system $|D'|$ is contained in the canonical series of D; in general, it will not be complete, and we denote by $\delta(D)$ its defect. By the Riemann-Roch theorem for curves, one deduces that the dimension of $|D|$ is given by

$$r' = p_g - 1 + \pi - \delta(D), \tag{34}$$

where π is the genus of D. Moreover, Enriques proved that if $|D_1|$

is another irreducible linear system contained in $|D|$, then $\delta(D_1) \leqslant \delta(D)$; in particular, $\delta(kD) \leqslant \delta((k+1)D)$ for all integers $k \geqslant 1$. Applying this to the system $|C|$ of plane sections of S, one finds from (34) that

$$\delta((k+1)C) - \delta(kC) = \omega_{n-3+k},$$

and, since in this case $\delta(C) = \omega_{n-3}$, it follows that

$$\delta(kC) = \omega_{n-3} + \omega_{n-2} + \cdots = q \tag{35}$$

for sufficiently large values of k. Finally, for *any* irreducible linear system $|D|$ of curves on S, $|D|$ is contained in $|kC|$ and $|kC - D|$ is irreducible for k large enough; thus the invariant formulation of Enriques' theorem (33) is that, for any linear system $|D|$, one has $\delta(D) \leqslant q$, and the maximum value is attained for some $|D|$. In 1905 E. Picard proved that for irreducible systems one *always* has $\delta(D) = q$; from this it follows that if $\omega_{n-3} = 0$, then S is regular—a result already proved by Castelnuovo [1897].

For an irreducible system of curves $|D|$ on S, one can also consider the *characteristic series*, defined—as for linear systems of plane curves—as the series cut on a generic curve of $|D|$ by the other curves of the system. Here the characteristic series will not necessarily be complete; by an argument similar to his method of deriving (33) (a construction of surfaces cutting given systems of curves on the planes of a pencil), Castelnuovo [1897] proved that the *defect* of the characteristic series is also at most the irregularity q, and that this maximum is attained.

A fundamental problem for linear systems of curves on a surface S was to evaluate its dimension—the generalization of Riemann's problem for curves. Enriques and Castelnuovo took the first steps in that direction; a linear system $|C|$ is said to be *special* if it is contained in the canonical system $|K|$ (if $|K|$ doesn't exist, every system is nonspecial); for a special system $|C|$, one writes $i - 1$ for the dimension of $|K - C|$, where i is called the *index of speciality* (for nonspecial systems, $i = 1$). When $p_g > 0$, the characteristic series of a complete system is always special; applying this to the defect of a characteristic series, Castelnuovo [1897] obtained the

"Riemann-Roch" inequality for the dimension r of a complete irreducible linear system of genus π and degree n:

$$r \geqslant n - \pi + p_a + 1 - i. \tag{36}$$

(This result had already appeared in [Noether, 1886] and [Enriques, 1896] for the case when $q = 0$.)

The most interesting contribution of Enriques is probably the introduction of new linear systems naturally associated to an algebraic surface S. It may happen that $p_g = 0$, but that the adjoint $|C'|$ of any linear system $|C|$ exists; then $|C|$ is not contained in $|C'|$, but it may be the case that for an integer $k > 2$ the system $|kC|$ is contained in its adjoint $|kC'|$. Then the system $|kC' - kC|$ is independent of the chosen system $|C|$; it is called the *k-canonical system*, and its dimension P_k is the *k-plurigenus* of S. If S has only ordinary singularities, the k-canonical system is cut by adjoint surfaces of S of degree $k(n - 4)$, which have the double curve of S as a multiple curve of multiplicity k.

The first example found by Enriques [1894] of a surface having genera $p_g = p_a = 0$ and a bigenus $P_2 = 1$ is the sextic surface (later called the *Enriques surface*) having the edges of a tetrahedron as a double curve. Another example was found by Castelnuovo: consider in $P_3(\mathbf{C})$ a conic, a line, and three points in general position; then the surface has degree 7, the conic is a double curve, the line is a triple curve, and, finally, at each of the 3 points the surface has a tacnode with tangent planes containing the triple line; then $p_g = p_a = 0$ and $P_2 = 2$.

Already in 1894, the concept of bigenus found a beautiful application in a paper by Castelnuovo [1896]. At first he believed that the conditions $p_g = p_a = 0$ were sufficient for a surface to be *rational*; the example of Enriques' surface led him to prove a remarkable result, that $p_a = 0$ and $P_2 = 0$ (which imply $p_g = 0$) give the rationality condition. The k-canonical system could also be written $|C^{(k)} - C|$, where $|C^{(k)}|$ is the *kth iterated adjoint* of $|C|$; Castelnuovo's method was used to investigate the properties of the sequence of iterated adjoints under the assumption $p_g = p_a = P_2 = 0$.

In another beautiful paper on the genus p of a curve of degree n in $P_3(\mathbf{C})$, Castelnuovo [1893] considered the linear series $g_{kn}^{r_k}$ cut on the curve by all surfaces of degree k; using the device of comparing r_k and r_{k-1}—similar to the one he used later for linear systems of surfaces—he proved that

$$r_k - r_{k-1} \begin{cases} \geqslant 2k+1 & \text{if } k < (n-1)/2, \\ \geqslant n & \text{if } k \geqslant (n-1)/2. \end{cases} \tag{37}$$

Then he proved—by contradiction—that if χ is the integral part of $(n-3)/2$, the series $g_{n\chi}^{r_\chi}$ cannot be special; using relation (37) and the Riemann-Roch theorem for curves, he proved that the defect d_χ of that series must satisfy

$$d_\chi \leqslant \chi(n-2-\chi) - p,$$

and, as $d_\chi \geqslant 0$, one must have

$$p \leqslant \chi(n-2-\chi). \tag{38}$$

In the preceding account, one of the main sources of difficulties encountered by Castelnuovo and Enriques was, for the sake of simplicity, suppressed. In contrast with the situation for curves, two smooth models of an algebraic surface are not necessarily isomorphic; for example, there are many smooth rational surfaces not isomorphic to the plane—the simplest one is the quadric. What happens is that for a birational transformation T of a surface S onto a surface S', there may exist curves on S (or S') that correspond by T^{-1} (or T) to points of S' (or S). These are called *exceptional curves*, and they (or the corresponding points) may be *singular*, even if the corresponding points (or curves) are not. So, one may well imagine the additional complexity this brought to the definition of linear systems of curves, which had to take these phenomena into account; indeed, it is almost impossible to check the correctness of the proofs involving them (see, for example, [Zariski, 1934]).

In a remarkable joint paper, Castelnuovo and Enriques [1900] limited themselves—as do modern algebraic geometers—to *smooth* surfaces (in some high dimensional space); the *adjoint* $|C'|$ of a linear system $|C|$ can be defined without reference to a projective embedding, by the condition that the curves of $|C'|$ cut on a generic curve $|C|$ the canonical series. The paper is rich in many new results, in particular on the new definitions of "relative" invariants such as the "Curvengeschlecht" of Noether and the "Zeuthen-Segre" invariant. But the most important part of their paper concerns the exceptional curves on a smooth surface. They showed that on such a surface S the following properties are equivalent:

1. S is birationally equivalent to a *ruled surface*.
2. There exists on S a linear system for which the degree m and the genus π satisfy $m > 2\pi - 2$.
3. S contains infinitely many exceptional curves.
4. The sequence $|C|$, $|C'|$, $|C''|$,... of successive adjoints of a linear system $|C|$ cannot be infinite—i.e., one reaches an iterated adjoint system that has no adjoint.

Furthermore, for smooth surfaces *not* birationally equivalent to ruled surfaces, Castelnuovo and Enriques showed that it is always possible to transform such surfaces into smooth surfaces having *no exceptional curves* (these were later called *minimal* models).

CLASSICAL ALGEBRAIC GEOMETRY AFTER 1900

Many things happened in the years immediately following 1900: Another first-rate geometer, F. Severi, appeared on the scene in Italy. Next, moving beyond linear systems of curves on a surface, the study of general *algebraic* systems—called "continuous systems" by the Italians—was undertaken in earnest; before 1900, the only nonlinear systems of curves which had been considered were the *irrational pencils* (an example of which is given by the generators of a ruled surface). The "transcendental" methods of E. Picard and G. Humbert, which hitherto had provided only alternate proofs to the geometric methods using linear systems, now took the lead with the introduction of new invariants linked to the theory of algebraic systems; this was crowned by a theorem of H. Poincaré for which no algebraic proof could be found before 1960

[Mumford, 1966]. It is also during this period that the relations between birational invariants and *topological* invariants of a surface were brought to light, before becoming the central theme of the work of S. Lefschetz. Until 1906 Castelnuovo took an active part in these developments and contributed some key results to the theory before leaving algebraic geometry. Enriques, however, continued his work in this area until the end of his life, arriving at his famous classification of algebraic surfaces—begun by him and Castelnuovo (see [Beauville, 1978]). Finally, the Italian geometers, and later Lefschetz, began the study of algebraic varieties of arbitrary dimension.

REFERENCES

Baker, H. F., 1933, *Principles of Geometry*, vol. 5, 6, Cambridge University Press.

Beauville, A., 1978, "Surfaces algébriques complexes," *Astérique*, 54.

Bertini, E., 1882, "Sui sistemi lineari," *Rend. Ist. Lomb.*, (2) 15.

Brill, A., and M. Noether, 1873, "Üeber die algebraischen Funktionen and ihre Anwendung in der Geometrie," *Math. Annalen*, 7.

Castelnuovo, G., 1892, "Ricerche generale sopra i sistemi lineari di curve piane," *Mem. Accad. Sci. Torino*, (2) 42.

――――, 1893, "Sui multipli di una serie lineare di gruppi appartenenti ad una curva algebrica," *Rend. Circ. mat. Palermo*, 7.

――――, 1896, "Sulle superficie di genere zero," *Mem. Soc. Ital. Sci. detta dei* 40, (3) 10.

――――, 1896, "Alcuni risultati sui sistemi lineari di curve appartenenti ad una superficie algebrica," *Mem. Soc. Ital. Sci. detta dei* 40, (3) 10.

――――, 1897, "Alcuna proprietá fondamentali dei sistemi lineari di curve tracciate sopra una superficie algebrica," *Ann. di Mat.*, 25.

Castelnuovo, G., and E. Enriques, 1900, "Sopra alcuni questioni fondamentali nella teoria della superficie algebriche," *Ann. di Mat.*, (3) 6.

Clebsch, A., 1868, "Sur les surfaces algébriques," *C. R. Acad. Sci.*, 67.

Enriques, F., 1894, "Ricerche di geometria sulle superficie algebriche," *Mem. Accad. Sci. Torino*, (2) 44.

――――, 1896, "Introduzione alla geometria sopra le superficie algebriche," *Mem. Soc. Ital. Sci. detta dei* 40, (3) 10.

Mumford, D., 1966, "Lectures on curves on an algebraic surface," *Ann. of Math. Studies*, 59.

Noether, M., 1870, 1875, "Zur Theorie des eindeutigen Entsprechende algebraischen Gebilde," *Math. Annalen*, 2, 8.

――――, 1886, "Extension du théorème de Riemann-Roch aux surfaces algébriques," *C. R. Acad. Sci.*, 103.

Picard, E., and G. Simart, 1897–1903, *Théorie des fonctions algébriques de deux variables indépendantes*, 2 vol., Gauthier-Villars, Paris.

Segre, C., 1887, 1889, "Recherches générales sur les courbes et les surfaces réglées algébriques," *Math. Annalen*, 30, 34.

Walker, R., 1935, "Reduction of singularities of an algebraic surface," *Ann. of Math.*, 30.

Zariski, O., 1934, *Algebraic Surfaces*, Springer, Berlin.

Zeuthen, H., 1871, "Etudes géométriques de quelques-unes des propriétés de deux surfaces dont les points se correspondent un à un," *Math. Annalen*, 4.

INDEX